Gaston Planté

The Storage of Electrical Energy

Gaston Planté

The Storage of Electrical Energy

ISBN/EAN: 9783744666596

Printed in Europe, USA, Canada, Australia, Japan

Cover: Foto ©berggeist007 / pixelio.de

More available books at **www.hansebooks.com**

THE STORAGE
—OF—
ELECTRICAL ENERGY

THE STORAGE OF ELECTRICAL ENERGY

AND RESEARCHES IN THE

EFFECTS CREATED BY CURRENTS COMBINING QUANTITY

WITH HIGH TENSION,

—BY—

GASTON PLANTÉ,

LICENCIÉ ÈS-SCIENCES PHYSIQUES,
ANCIEN PROFESSEUR DE PHYSIQUE À L'ASSOCIATION POLYTECHNIQUE
ETC.,

FROM
1859 TO 1879.

WITH EIGHTY-NINE ILLUSTRATIONS.

Translated from the French by

PAUL BEDFORD ELWELL,

(OF ELWELL-PARKER, LD.)

LONDON:
WHITTAKER & CO.,
2, WHITE HART STREET, PATERNOSTER SQUARE, E.C

A Sa Majesté

DON PEDRO D'ALCANTARA

Empereur du Brésil,

Associé étranger de l'Académie des sciences de l'Institut de France.

Sire,

Je prie Votre Majesté de daigner agréer la dédicace de ce livre comme un faible témoignage de ma profonde reconnaissance.

Vous avez été le premier à encourager mes travaux. Après avoir assisté, en 1872, à mes expériences sur les courants secondaires, au Conservatoire des Arts-et-Métiers, Votre Majesté a bien voulu, en 1877, honorer deux fois de sa visite mon laboratoire de la rue de la Cerisaie, dans ce quartier du vieux Paris, où les noms d'Henri IV et de Sully sont encore vivants.

Cent soixante ans auparavant, Pierre le Grand venait y habiter l'hôtel Lesdiguières, et, vers la fin du dernier siècle, l'illustre Franklin assistait, dans l'hôpital des Célestins, à des expériences d'électricité.

La présence, dans ces mêmes lieux, de Votre Majesté poursuivant son enquête sur tous les progrès utiles à l'humanité, ajoutera une nouvelle page à nos anciennes traditions et un précieux souvenir qui sera conservé.

Je suis,

avec le plus profond respect,

Sire,

de Votre Majesté,

le très-humble et très-obéissant serviteur,

GASTON PLANTÉ.

ERRATA.

PAGE 14—Line 11 from bottom, *after* "like" *insert* "the oxides of."

" 18———5, *after* "two cells" *insert* "acting on a voltameter with wires."

" 22———5, *for* "Dewar" *read* "Dumas."

" 48———9, *for* "production" *read* "reduction."

" 53———14, *for* "Bollot" *read* "Boillot."

" 53———16, *for* "positive" *read* "negative."

" 53———19, *for* "production" *read* "reduction."

" 69———5, *for* "also used in connection" *read* "compared."

" 69———8 from bottom, *for* "powerful" *read* "potential."

" 75———4, *for* "Thomson" *read* "Thomsen."

" 111———12, *for* "Willigan" *read* "Willigen."

" 126———7, *for* "Sheath" *read* "Sheaf."

" 37———3 from bottom, *for* "combination of lime and silica" *read* "lime combined with silica."

" 138———3 from bottom, *for* "silica" *read* "silicium."

" 141———2 of the note, *for* "p. 58" *read* "p. 44."

" 148———7 of the note, *for* "rorid" *read* "roric."

" 186———16, *for* "Brissou" *read* "Brisson."

" 209———11, *for* "supplied by" *read* "in connection with."

" 221———4 from bottom, *for* "fig. 80" *read* "fig. 79."

" 223———13, *for* "fig. 79" *read* "fig. 80."

" 238———6 of the note, *for* "Vol. 1" *read* "p. 214."

" 257———11, *read* "when there is a break or change in the material."

PREFACE.

This work includes the main results of the researches we have contributed to the Académie des Sciences, or published in various scientific periodicals, from the year 1859 to 1879.

It is divided into six parts.

The first comprises a description of the experiments and apparatus we have made known for *accumulating* or *transforming* the energy of the voltaic battery by means of secondary currents.

The second part contains an enumeration of the applications to which these experiments have been put, and of others which might be carried out.

The third part relates to phenomena observed with electric currents of high tension obtained by the means described in the first part.

The fourth part treats of analogies which these effects seem to present with many great natural phenomena, and the inferences drawn from them in order to explain these phenomena.

The fifth part embraces the description and study of the effects of a new apparatus, by means of which we have tried to transform, in the most complete manner possible, dynamic into static electricity, and which we have distinguished by the name of *rheostatic machine*.

The sixth is devoted to a concise enumeration of analogies which the electrical phenomena (particularly those which we have noticed with currents of very high tension) present with effects produced by mechanical actions, and to the inferences drawn from them as to the nature of electricity.

The reader who will only admit absolute deductions from facts, may omit the fourth part, in which induction has considerable play.

Nevertheless, we thought we ought not to pass over in silence some of the ideas which the results of our experiments presented, and the real or apparent analogy which they show with natural phenomena.

Authors are often blamed for not having understood the inference to be drawn from facts which they have observed, or for not having perceived the theoretic sequence, and the applications to which it may lead. We have sought to be free from this fault, which, besides, has not always foundation; for it is rare for anyone who has patiently studied nature, seeking new facts, not to also meditate upon their significance, and, as nature is seen complete in each one of her manifestations, it is difficult for the student to avoid being led to generalise the result of his observations. This tendency becomes, no doubt, another reef upon which one may strike; yet, science would not lose, we think, by these generalisations, or by the hypotheses to which they lead, from the moment they rest no longer upon pure imagination, but are inspired by attentive observation of facts, and are set forth, besides, with reserve, without elevating them as doctrine, without affirming that they are true.

This is what we have tried to do, and, to follow the example of one of the greatest thinkers of past ages, we will humbly say, in publishing these researches : " *Quæro, Pater, non affirmo.*"[1]

---

(1) St. Augustine.

FIRST PART.

The Accumulation and Transformation of the energy of the Voltaic Battery by means of Secondary Currents.

CHAPTER I.

Secondary Currents.—Voltaic Polarisation.—Study of the Secondary Currents produced by different Voltameters.—Conclusions.

1. SECONDARY CURRENTS were observed at the beginning of this century, shortly after the discovery of the Voltaic Battery. Gautherot, a French Scientist, was the first to discover, in 1801,[1] that platinum or silver wires which had been used to decompose saline water by this battery, possessed the property, after having been cut off from the battery itself, of giving an electric current of short duration.

Ritter[2] made the same discovery at Iéna, with gold wire, and made the first secondary battery, by superposing a series of pieces

---
(1) Memoirs of the Learned and Literary Societies of the French Republic, 1801.
(2) See Exposés des Travaux de Ritter par Œrsted. Journal de Physique 1803. t. LVII, P. 345.

of gold, separated by cloth discs, moistened with a saline solution. This battery, inactive in itself, after having been submitted to the action of a voltaic battery of a greater number of cells than that of which it was composed, could give for some moments a current in the opposite direction to that of the voltaic battery. This current took the name of secondary current.

Ritter varied the kind of metal, and the number and surface of the plates composing the secondary battery.

He used platinum, copper, brass, iron, and bismuth, and found that gold, platinum, and silver, gave a stronger secondary current than any of the other metals. He noticed that carburet of iron, and peroxide of manganese, gave still more marked results; but he obtained no effect with lead, tin, and zinc.[1]

The Secondary Batteries that Ritter employed more especially in his experiments, were formed of discs of copper separated by circular pieces of cloth, moistened in saline water, or sal amoniac. By charging a secondary battery formed of a column of fifty copper discs, by means of a zinc and copper voltaic battery, of a hundred pairs, he obtained decomposition of water, various chemical or physiological actions, and, in general, all the effects produced by ordinary batteries. Nevertheless Ritter's Secondary Batteries, in consequence of the objections arising from their arrangement in a columnar form, or crown of cups, like that given to the voltaic battery itself at that period, in consequence also of the brief duration of the currents produced, and the necessity of employing a battery of a greater number of clements to charge

---

. [1] We shall explain further on (22) why Ritter obtained no result with lead, the metal which we have on the contrary exclusively used for obtaining powerful Secondary Currents.

them, could not be advantageously applied either in scientific research or commerce.

2. The secondary current produced by one of Ritter's batteries has been explained by Volta, Marianini, and Becquerel, who have shown that this current arose from the formation of acid and basic deposits upon the metal discs, in consequence of decomposition of the salt impregnating the moistened pieces of cloths, under the influence of the primary current. Becquerel in particular proved clearly the production of an electric current by the reciprocal action of an acid and its base, by surrounding two platinum plates, one with an acid, the other with a basic solution, united by a moist conducting material. He thus formed a battery element of which the plate immersed in the acid solution, constituted the positive pole, and the plate in the basic solution formed the negative pole. The direction of the current of the secondary battery became thus explained, this direction being such that the disc in connection with the positive pole of the primary battery, is itself the positive pole of the secondary battery.

3. About 1826, attention was again called by de la Rive to the secondary current arising from platinum plates in a voltameter filled, not with a salt solution, but with water slightly acidulated by sulphuric acid, or even with distilled water.

As the presence of acid or basic deposits was out of the question in this case, the secondary current was at first attributed simply to physical effect produced by the primary current, to a special polarization of the plates under its influence, and this current received the name of *Polarisation Current*.

4. This term has remained in scientific use; but it has since been found that the gases developed around the plates, even in a

very small quantity, must be the cause of the current; for Matteucci obtained a current with platinum plates, previously immersed in oxygen and hydrogen gas, and Mr. Grove made a gas battery, by coupling up a certain number of elements formed of platinum plates, immersed at one and the same time in acidulated water, and inverted test tubes of oxygen and hydrogen. This battery was not of course very strong, nevertheless decomposition of water could be obtained by it, and consequently it presented an interesting voltaic circle, the synthesis and analysis of water being exhibited at the same time, in one experiment: the synthesis producing the voltaic current, shown by the gradual absorption of the gas in the test tubes; the analysis produced by the passage of the current in the exterior circuit, demonstrated by the liberation of oxygen and hydrogen gas in a voltameter.

5. Voltaic Polarization has since been the object of interesting labours on the part of a great number of scientists, notably Faraday, Wheatstone, Schœnbein, Poggendorff, Buff, von Beetz, Svanberg, Lenz and Saweljew, Edmund Becquerel, du Moncel, Gaugain, &c.

The object of these works has been, in general, studying or measuring the secondary current obtained with platinum electrodes.

We remember, however, that Mr. Sinsteden,[1] in a memoire upon the results obtained with a magneto-electric apparatus, having by haphasard sent the current from this apparatus through voltameters with lead, silver, and nickel plates, obtained secondary currents with these metals, sufficiently intense to raise wires to a state of incandescence.

---

(1) Sinsteden. Recherches sur le degré de force et de continuité du courant d'un grand appareil magnéto-électrique de rotation. Annals de Poggendorff. t. LXXXXII, p. 16, 1854.

6. **The study of secondary currents produced by various voltameters.**—The researches we have, on our part, undertaken in this subject, in 1859,[1] were made with the special object of comparing the secondary currents produced by voltameters of various metals in different solutions, by taking observations directly after breaking the primary circuit, and at the same time, studying the details of the phenomena manisfested in these voltameters, with regard to the development of a secondary current.

Figure 1 shows the voltameter or voltascope used. This apparatus was arranged so as to enable the effects produced around the electrodes to be easily followed (*Fig. 1*). We took care to light it well, and its transparency admitted of perfect observation. It was arranged to take four wires, for studying, in certain cases, the part taken by each electrode in producing the secondary current.

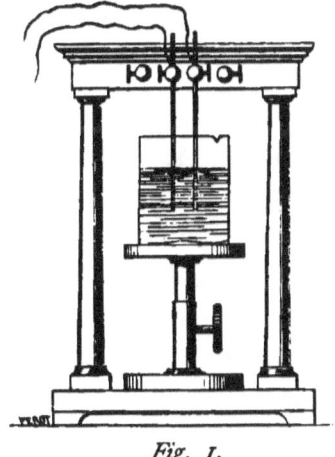

Fig. 1.

All observations were taken under the same conditions, by employing wires of one diameter, immersed in the same quantity of liquid, placed an equal distance apart, and subjected for the same time to the action of the primary current.

A mercurial commutator (*Fig. 2*), similar to that of Ampère, closes the secondary circuit through SS', the instant the primary curcuit PP' is broken.

---

(1) Recherches sur la polarisation voltaïque. Comptes-rendus, t. XLIX, p. 402, 1859. Bibliotheque universelle de Genève, t. VII, p. 292, 1860.

Fresh researches upon polarisation were again made since this period, by Messrs. du Bois-Reymond, Crova, Raoult, Thomson, Paalzow, Patry, Parnell, Branly, Thayer, Lippmann, Bernstein, Fleming, Colley, Hankel, etc,

Fig. 4 shows the general arrangement of the apparatus.

P is the primary battery, composed of from two to four Bunsen cells, *I* is the commutator, *V* the voltascope.

Fig. 2.

At *G*, is a galvanometer, or tangent compass, allowing the variations in intensity of the primary current to be followed, as polarisation is developed in the voltameter.

The maximum polarisation was obtained rather rapidly, as the wires only presented a very small surface, and the resistance of the galvanometer was very low.

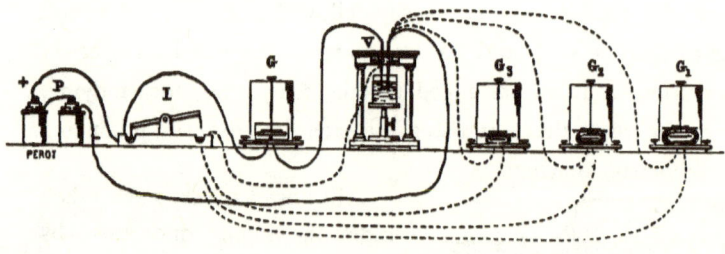

Fig. 3.

At G^1, G^2, G^3, were placed three galvanometers of different degrees of sensitiveness, for taking observations of the secondary current. One of these galvanometers was replaced by a tangent compass, when it was desired to measure the intensity, and when the indications of the other two galvanometers had given an approximate idea of the power of the secondary current. This current only possessing, in certain cases, a very short duration, the deviation was noted by placing beforehand a stop behind the needle, at a point very close upon the maximum degree which it could attain, which was determined approximately by several preliminary experiments. This stop was then moved forward until the needle could only make a slight movement of about half a degree further, after closing the secondary circuit.

7. We have thus recognised that the order in which metals used as electrodes in a voltameter, with water acidulated by sulphuric acid, could be classed, with a regard to the intensity of the secondary current, observed immediately after breaking the primary current, was not exactly the reverse of the order in which these same metals were classed with regard to the intensity of the primary current passing through the voltameter, such as would have been the case, if the tendency to the production of a secondary current, or the polarisation E.M.F. as generally termed, was the principal or only cause of the enfeebling of the primary current.

Thus, to give the most striking example, we have noticed that the voltameter with aluminium electrodes is the one which most weakens the intensity of the primary current, for it ends by almost completely arresting the passage of the current after a certain time.[1] This metal would be consequently placed last on the list of metals, classed in the order of intensity in the primary current passing through the voltameter. If the cause of the enfeebling of the primary current were only due to the secondary E.M.F., aluminium should give the strongest secondary current, and would consequently be placed first upon the list of metals classed according to the intensity of the secondary current which they give in the voltameter. Now, it is not thus, for aluminium is found to give a weaker secondary current than any other metal. This arises from the fact that other causes, often more influential than the secondary current itself, such as the formation of a layer

---

(1) Comptes-rendus, t. XLIX, p. 403, 1859.

of oxide at the positive pole, and its insolubility and want of conductivity, also the sheaths of liquid arising from the action of the metal upon the electrolyte, at one or other of the poles, contribute towards weakening the primary current.

This is the result of the careful examination we have made of the phenomena which take place in voltameters composed of various metals and different liquids.[2]

8. **Voltameter with copper electrodes.**—In a voltameter, copper presents a series of phenomena which clearly show the influence exercised by the oxidation of the metal at the positive pole, and by the more or less dissolving action of the liquid surrounding it upon the intensity of the primary current.

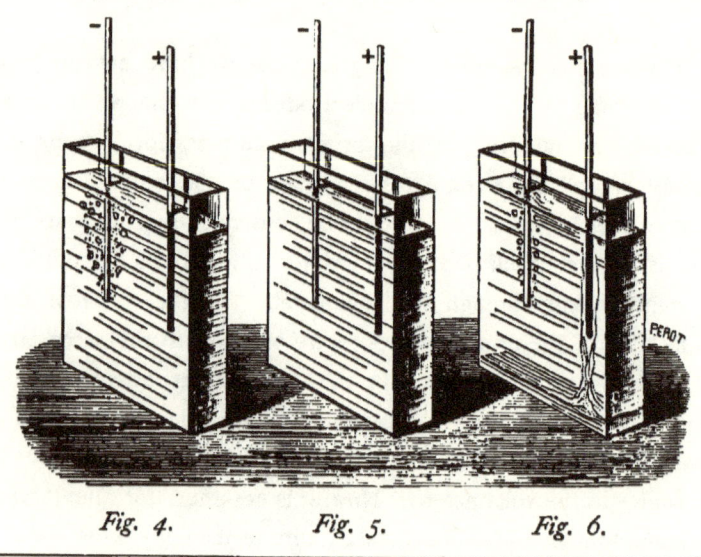

Fig. 4.   *Fig. 5.*   *Fig. 6.*

---

(2) M. Gaugain in a work upon the secondary E.M.F. of platinum plates in voltameters (Comptes-rendus, t. XLI, p. 1165, 1855), showed the interest there was in studying the effects of this force, after the electrolysis, that is to say, after the action of the primary current. M. Gaugain concluded that the diminution in intensity resulting from the presence of an electrolyte in a voltaic circuit was a phenomenon much more complex than had been hitherto supposed.

If the current from one or two Bunsen cells be passed through a voltameter with copper electrodes, in water acidulated by sulphuric acid, three distinct phases are observed:

First, immediately the connections are made, the positive wire becomes darkened without liberating any gas, the negative wire freely liberates hydrogen gas, and the galvanometer makes a strong deviation.

Secondly, the liberation of hydrogen decreases and is almost stopped for a moment. The needle of the galvanometer returns very nearly to zero. This phase corresponds to the formation of the full thickness in the layer of oxide of copper round the positive pole *(Fig. 5.)*;

Thirdly, the current being stopped, the oxide of copper begins to be dissolved; the positive wire is stripped of the greater part of the layer of oxide which envelopes it, and gives rise to a flow of sulphate of copper, in the form of a dense bluish liquid line, towards the lower part of the cell. At the same time the liberation of hydrogen recommences at the negative electrode; but it is less abundant than during the first period *(Fig. 6)*. The deviation of the galvanometer likewise increases, but it is far from attaining the original point.

The current having resumed a certain intensity, it appears that a fresh layer of oxide ought to be deposited upon the electrode, and then the phenomena of stopping and recommencing would take place indefinitely, so long as the passage of the primary current continued. But the layer of copper sulphate, formed upon the wire, produces a movement in flowing away which prevents the oxide adhering in a thick layer, and allows of its being dissolved to a great extent, as quickly as it is formed. The positive wire once stripped of the layer of oxide formed, apparently does not become recoated, and the copper continues to be dissolved without the preliminary phase of oxidation. The metal is only slightly discoloured.

The intensity of the current during the third period, is reduced to about one sixth of what it was during the first moments; consequently equilibrium results between the two actions which tend to be produced nearly simultaneously; namely, the formation of a bad conducting oxide, under the never ceasing influence of the current, and the solution of this oxide by the acidulated water in the voltameter.

Nevertheless, this proportion between the original and final intensity is not always maintained in the same degree; it varies according to the duration of the experiment; because, the liquid becoming gradually charged with copper sulphate, the negative wire becomes coated with reduced, spongy, copper; the conducting power of the voltameter increases in consequence of this deposit, and the intensity of the primary current naturally tends to increase.

9. The secondary current, tried immediately after each of these three phases, practically showed the same intensity; which shows that the remarkable variations in intensity in the primary current of which it has just been question, do not mainly depend upon the secondary current.

This current is besides very weak in a voltameter with copper electrodes and water acidulated with sulphuric acid. In order to measure the E.M.F., the same as in voltameters of some other metals, immediately after breaking the primary current, we have used Becquerel's electro-magnetic balance, by ascertaining, in a series of successive tests, the weight which it was necessary to place beforehand in the scales, in order that the secondary current might still produce a slight movement of the beam of the balance during the first moment of passing. We have thus found that the E.M.F. of the secondary current given by a voltameter with copper electrodes, and water acidulated by sulphuric acid, immediately after breaking the primary current, was only about one tenth that of a Daniell element.

10. If, during the passage of the primary current through a voltameter with copper electrodes and water acidulated with sulphuric acid, when the intensity is reduced to that of the third phase, the positive electrode be shaken, the liberation of hydrogen becomes more abundant at the other electrode; the primary current increases and keeps, to a certain extent, constant during the whole time that the movement continues.

The effect of this disturbance is to separate the layer of saline liquid which surrounds the wire and to favour the solution of oxide of copper by the acidulated water, as it becomes formed. This layer is then an important cause of the falling off in the primary current, and has more influence in the case in question than the tendency of the voltameter to produce a secondary current; for this latter current, far from being weaker after the disturbance of the voltameter, also increases in intensity. This fact is owing to the sulphate of copper being rapidly dispersed in the liquid of the voltameter during its disturbance; the negative electrode becomes coated as described previously (8) with a spongy deposit of reduced copper, and this wire, thus altered, is found to be in a favorable state for the production of a more intense secondary current, as we shall explain further on in detail (52), when treating of the chemical actions in secondary cells with lead plates.

As regards disturbance of the negative electrode in a copper voltameter, it has no effect; for it only displaces the liquid a little more (already set in motion by the bubbles of gas liberated round the wire), and the liquid does not otherwise exercise any action by itself upon the negative wire.

11. If, instead of water acidulated by sulphuric acid, we use concentrated sulphuric acid in a copper voltameter, the layer of oxide formed at the positive pole is not dissolved at all and preserves the underlying metal from any attack, and it remains unchanged under the influence of the current. Sulphate of copper ($CuO, SO^3, 5HO$) cannot be formed, there not being the necessary water in the concentrated sulphuric acid for its formation. The current is gradually reduced to zero, in consequence of the resistance of the oxide.

12. **Voltameter with silver electrodes.**—The phenomena observed in a copper voltameter are reproduced with a few modifications in a voltameter with silver electrodes. We observe the three phases, of maximum, minimum, and recommencement of the current, especially when only employing a single Bunsen cell as primary source.

The positive wire becomes a deep grey colour, or blackened; but the period of decrease only lasts a very short time, so that it is more difficult to measure than with copper. The positive wire is not completely stripped of the layer of oxide as quickly as it is formed; it remains covered in spite of the flowing away of the saline liquid.

The secondary current is much more intense than in the case of copper; but it is also of short duration, in consequence of the spontaneous solution in the liquid of the layer of oxide which constitutes the main source of the secondary current, as will be seen further on.

13. A phenomenon is shown with silver which is not perceived in the case of copper. When the primary current is cut off, the positive electrode which has preserved a thin deposit of oxide upon its surface, gives rise to a liberation of gas, even when the circuit is not closed. This liberation arises from local action taking place between the oxidised surface of the wire, and the metal underneath it. It is known that electro-chemical deposits, especially of this nature, are easily penetrated. The liquid then comes into contact with the metal and its oxide at the same time; the oxide soon disappears, and leaves visible the silver wire in a metallic state. If, at that time, the secondary circuit is closed, no current is noticed.

This fact proves that the secondary current must arise particularly from the chemical action produced at the positive electrode.

To confirm it, a third silver wire, not intended to have the primary current passed through it, is immersed in the voltameter. It is arranged so that at the moment the secondary circuit is closed, it becomes connected with the oxidised positive wire, whilst the negative wire remains out of the circuit. Now we observe that the secondary current exhibits nearly the same intensity as when the two electrodes are submitted to the action of the current. The negative electrode does not then contribute sensibly to the production of the secondary current.

14. The effects observed by causing a disturbance in a voltameter with silver electrodes, are the same as those presented in the copper voltameter. The dissolved silver salt is distributed in the solution, the negative wire becomes covered with a spongy, metallic deposit, and the secondary current becomes noticeably increased.

15. In a silver voltameter with concentrated sulphuric acid, the positive silver wire oxidises, and becomes gradually dissolved, the reverse of what takes place with copper in the same liquid (11). This is explained by the chemical composition of sulphate of silver (AgO,SO^3) in which combination water does not take part.

16. **Voltameter with tin electrodes.**—The three phases are very clearly marked in the case of this metal. The positive wire blackens, and gives rise to an abundant flow of *protoxide(?) sulphate* of tin. The secondary current, although less strong than that from silver, has nevertheless considerable intensity. Disturbance produces, as with the preceding metals, an increase of the primary current, a deposit of tin upon the negative electrode, and an increase in the secondary current.

17. **The lead wire voltameter.**—This voltameter does not show the three successive phases in the intensity of the primary current that is noticed with the preceding metals. This is because the peroxide formed round the positive pole is insoluable and coats the electrode in a permanent manner, without gradually disappearing in the solution like the metals already studied.

This adherence and insolubility of the peroxide of lead, added to its affinity for hydrogen on account of the high degree of its oxidation, contribute to produce, in a voltameter with lead electrodes, a more intense secondary current and longer in duration than that of any of the other metals.

18. Lead covered with peroxide of lead, in water acidulated by sulphuric acid, acts in fact in a manner exactly the reverse to that of zinc in the same liquid. It tends to decompose the water, by absorbing hydrogen, and to become the positive pole of a cell,

---

Query by Translator.—Stannous Sulphate.

if it is connected with lead not oxydised, whilst pure zinc tends to decompose the water, by absorbing oxygen, and becomes the negative pole of a cell in which it is opposed to another metal.

To this cause of the development of a secondary current by the voltameter of lead electrodes, we may add the effect produced upon the wire or plate of the negative pole on short-circuiting the voltameter after being submitted to the primary current.

The lead plate placed at the negative pole does not undergo, by the action of the primary current, as marked a change as that of the positive pole; nevertheless, as lead is always more or less oxydised by exposure to the air, it is brought to a more perfect metallic state by the hydrogen which is manifestly the means of reducing the cell, and its tint changes from bluish grey to a much lighter grey.

When we then close the secondary circuit, water being decomposed in the cell, simultaneously with the appearance of hydrogen upon the peroxidised plate, oxygen is carried to the plate rendered formerly metallic by the primary current, and oxidises it lightly. This oxidation is even visible; as the negative lead plate immediately darkens upon the closing of the secondary circuit. A single plate of lead,. or one united with another plate exactly the same, would be thus oxidised in water acidulated by sulphuric acid, and it would not give any E.M.F., no more than would pure or amalgamated zinc. But as the union of pure or amalgamated zinc with another metal less easily attacked, both immersed in acidulated water, or better still, in a liquid that can combine with hydrogen, occasions the zinc to be attacked, and consequently the development of a current; so the union of an ordinary lead plate with a peroxidised lead plate, which tends to decompose water by absorbing hydrogen, induces at the same time, oxidation of the other plate, and consequently the developement of an extra E.M.F. arising from this oxidation.

19. Such is the double chemical action which takes place in a voltameter of lead electrodes, upon the closing of the secondary circuit after breaking the circuit of the primary current. And such is the double cause of the development of the strong secondary current afforded by this metal.

If the circuit of the secondary cell be not closed after breaking the primary circuit, there is produced nevertheless, a visible chemical reaction in the voltameter, like that we have already described in the case of silver (13). There is a slight liberation of gas for some seconds at the positive plate. This fact is explained, in like manner, by the affinity of peroxide of lead for hydrogen, which causes local action to take place between the oxidised surface and the metal beneath, and, consequently, decomposition of the solution.

The action of metallic peroxides in absorbing hydrogen carried from the electro negative element, and increasing the E.M.F. of primary batteries, had been also noticed by Becquerel in regard to peroxide of manganese, and De la Rive had obtained, by means of powdered peroxide of lead, heaped around platinum or carbon, a cell of a higher intensity than either Grove's or Bunsen's.[1]

20. We have measured, several times, the E.M.F. of a voltameter with lead electrodes completely polarised by a sufficiently prolonged action of the primary current. We have found, by operating immediately after breaking the circuit of this

---

[1] Archives de l'Electricité, t. III, p. 159, 1843.

current, that it was approximately equal to 1·5, a Bunsen element being taken as a unit.[1]

This E.M.F. has since been measured by M. Edmond Becquerel,[2] during the actual passage of the primary current, so as to determine, by the difference, the fall caused in the E.M.F. of the battery by the interposition of the voltameter with lead electrodes. This E.M.F. was thus found equal to 1·41, the current from the Bunsen cell being 1.

21. The disturbance of one or other of the electrodes in a voltameter of lead wires or plates does not produce the effects observed with the preceding voltameters, which is easily explained if we consider that the effects are due to the separation of the liquid in different degrees of density, arising from the decomposition of the various metals, and that the lead remains coated with an insoluble oxide in the liquid in which it is immersed.

22. If we use salt water in the voltameter, instead of water acidulated by sulphuric acid, there is formed round the negative pole chloride of lead, scarcely soluble, and a very bad conductor, so that the primary current is rapidly diminished, and the secondary current itself is much weaker than that which is produced in water acidulated by sulphuric acid. We have found that the E.M.F. of this current measured, as in the preceding cases, immediately after breaking the primary circuit, was but ·08 that of a Bunsen element.

This result explains why Ritter, who nearly always used salt water for solution, did not notice any marked effect with lead, and did not employ this metal in his secondary batteries.

---

(1) Comptes rendus, vol. 50, p. 640, March 1860.
(2) Annales du Conservatoire, No. 2, p. 277, October 1860.

23. Voltameter with aluminium wires.—Aluminium shows, in a marked manner, the failing which takes place in the primary current by reason of the insolubility and resistance of oxide formed in the voltameter.[1]

Two cells of this metal give a strong deviation; but it quickly disappears, and there remains but a very feeble current.

It is even the same with four cells. It may be easily proved that this diminution is owing to the bad conductivity and insolubility of the oxide.

First, we show that, when the current has become very weak, if the circuit be broken for a few moments, and then closed again, the current does not regain its original intensity.

That being so, if we replace the positive wire by a freshly brightened or new wire, a very decided deviation will be observed, which disappears in its turn.

Changing the negative wire only causes a hardly noticeable increase of current.

22. The positive wire, taken out and examined immediately upon the decrease of current, shows no change in color or appearance, although it is probably oxidised. If it be washed, thoroughly wiped and replaced in the voltameter, it does not allow the current to pass any more easily; it must be rubbed with glass paper before a momentary but decided deviation is observed.

Shaking has no effect upon either wire.

The secondary current is extremely weak with this metal, so much so, that the great decrease in the primary current cannot be attributed to the opposing E.M.F. of the secondary current.

---

[1] Comptes rendus, vol. XLIX, p. 403. (1859), and Bibl-univ; de Genève, vol. VII, p. 292, 1860.

24. Voltameter of iron and zinc wires.—We have not spoken until now of the primary current which may be given by the voltameter, because the preceding metals are not sensibly attacked in acidulated water. Yet, when both wires are very well brightened, they sometimes give a slight deviation with a sensitive galvanometer, but this effect is not worthy of notice compared with those we have taken into consideration.

With iron and zinc, which decompose rapidly in acidulated water, the least difference in the attack of the two wires gives a current strong enough to be worthy of attention; besides, the direction of the current changes every minute, according to the degree in which each wire is attacked. We cannot then easily determine either the quantity or the E.M.F. of the secondary currents which are produced. There is however a means of observation which we may use, in order that the effects may be as little disturbed as possible by the primary current of the voltameter; it is to abstain from dipping the electrodes into the liquid beforehand, and to arrange the connections so that the circuit may be closed by the act of dipping the electrodes into the acidulated water. By operating in this manner, the electrodes become immediately submitted to the action of the primary current, without having undergone any preliminary attack by the solution itself.

We then find, more especially with iron, the three phases of intensity in the primary current, which the greater number of metals present. The positive electrode is darkened, and produces an issue of sulphate of iron, or of zinc, without liberating any gas. But upon breaking the circuit, the solution acts freely upon the metal, which immediately discharges hydrogen, and with so much the more energy as its surface remains slightly oxidised. The negative electrode on the contrary, having attained a more metallic state, under the influence of the electrolytic hydrogen, produced by the primary current, remains some time in the body of the acidulated water, without liberating any gas; but little by little it necessarily ends by becoming attacked. We may consider that the electrolytic hydrogen, by enveloping the metal, and preventing its oxidation by the solution during the passage of the primary current, plays the same part as mercury does in amalgamated zinc.

The secondary current, tried by working as we have described, is rather strong, and may be taken, during the first few moments, as almost entirely due to the action of the primary current, but the attack upon the electrodes by the solution itself soon disturbs the effects.

25. **Voltameter with gold electrodes.**—The positive gold electrode becomes visibly oxidised under the action of a rather weak primary current (2 Bunsen elements), and becomes coated with a reddish deposit of oxide of gold. If we change the direction of the current, the oxide is rapidly reduced by the hydrogen; the electrode becomes blackened without showing the metallic lustre. Another change in the direction of the current causes a fresh oxidation, still more rapid, on account of the finely divided state of the metal, but it is no longer possible to tell by its appearance, whether it is oxidised or reduced.

The secondary current is rather strong, but not very susceptible of an exact measure, because of its short duration. Unlike what happens with the other metals, it appears due, to a great extent, in spite of the decided oxidation of the positive electrode, to the action of hydrogen on the negative electrode. We recognise it in working as described previously (13), that is to say, by placing beforehand in the voltameter, a third and a fourth electrode outside the principal circuit. These electrodes being successively coupled with the positive and negative electrodes, immediately after the action of the primary current, we find that the secondary current given by the negative electrode and one of these nonpolarised electrodes, is stronger than the current produced by the positive and the other nonpolarised electrode. We explain this fact further on (27, 28, 29,) when treating of the platinum voltameter giving the same result.

26. **Platinum Wire Voltameter.** — The phenomena produced by the previously studied voltameters may help to account for those which are produced by the platinum wire voltameter, the explanation of which is also not devoid of difficulties.

If the primary current is supplied by a single Grove or Bunsen cell we know that electrolysis cannot take place, and that the current is almost completely reduced to zero by putting the voltameter in the circuit. If the primary current is supplied by two cells, electrolysis takes place, and the current, after shewing some decrease, is maintained constant without first going nearly to zero, as happens with metals easily oxodised. Platinum does not visibly oxidise at the positive pole, although the secondary current produced is strong; it is decidedly stronger than that given by the gold wire voltameter, but like the latter, it is of too short a duration for anyone to take an exact measure of the E. M. F. directly after the breaking of the primary circuit.

27. It proves, the same as with gold, that the action of the wire which evolves hydrogen during the electrolysis helps much more

to produce the secondary current than that of the wire which evolves oxygen.

28. This fact is explained by the condensation—or even combination of the hydrogen gas with the platinum,—which seems to behave like a metallic vapour, as Prof. Dewar has drawn attention to long ago. This combination being of course extremely transient and easily oxidised, we see that it may become the source of an electric current, and form the negative pole of a battery element, in relation with another wire of the same metal which has not undergone a similar action.[1]

Again the action of the electrolytic gases does not act only upon the metal electrodes but also upon the solutions in the immediate neighbourhood of the electrodes, producing some very unstable combinations with the liquid itself. The oxygen liberated round the positive platinum electrode in water acidulated with sulphuric acid forms ozone from the oxygenised water, and as recently shown by M. Berthelot,[2] a new compound, per-sulphuric acid, the discovery of which we owe to him.

These various bodies, all highly charged with oxygen, must contribute towards the partial production of the secondary current by the positive electrode.

29. The hydrogen liberated around the negative electrode forms on the other hand a compound corresponding to the oxygenised water, that is to say, a body more highly charged with

---

(1) Graham's Works have shewn, as is known, this tendency of Hydrogen to ally itself with metals, especially in the case of palladium; Prof. Dewar has observed a strong secondary current with a plate of hydrogenised palladium in connection with an ordinary plate of the same metal, and the researches of Messrs. Crova, Root, &c., have proved that a similar effect was possible with platinum.

(2) Comptes, Rendus, Vol. 86, pp. 20, 75, 1878.

hydrogen than the water itself, such as a sub oxide or protoxide of hydrogen, which would further contribute towards the production of the secondary current. Although a body of this kind has not yet been set free, the strongly reducing or oxidising tendencies of the electrolytic gases and the chemical symmetry of the phenomena which take place at the two poles of the cell, permit of the idea that there is a partial combination of the gas with the liquid round the negative electrode, and in consequence the formation of a compound of this nature.

30. There may also be another cause for the part taken by the positive pole in the developement of the secondary current by platinum electrodes, which there is occasion to mention: it is the oxidation to a very feeble extent of the platinum itself by the oxygen of the voltameter.

This oxidation, it is true, is not visible, but as it is so very evidently produced on all metals—even gold itself—by the active oxygen from the electrolysis, we are naturally led to think that platinum cannot be completely free from this strong oxidising effect.

It was the opinion of De La Rive, who, after having first admitted a simple physical action of the gases upon the electrodes, afterwards thought that there must be chemical action produced upon the platinum.

He had even observed the deterioration of electrodes of this metal which had served for the passage of a current in a voltameter alternately in both directions, and had rightly concluded that platinum, in order to arrive at this state, must have undergone successive periods of oxidation and reduction.

The fact given previously (23), as to aluminium not changing in appearance at the positive pole, although it oxidises to such an extent as to entirely stop the passage of the current, proves, moreover, that a metal may be coated with a very thin layer of oxide without altering its appearance.

A further proof in support of the fact of a very thin coat of oxide being formed upon the positive pole of the platinum voltameter is shewn by a platinum positive plate, washed and dried, without, of course, being rubbed hard, preserving the power of giving a secondary current when coupled with an ordinary plate, the same as the positive electrode of an aluminium voltameter, likewise washed and dried, retains the property of very nearly stopping the passage of the current, on account of the very thin, invisible coat of oxide with which it is covered.

Thus, the chemical change in platinum, produced even to a very feeble extent by the primary current, can assist, in some degree, towards the developement of a secondary current.

31. We may finally add to the above causes the action upon the electrodes of the gases in the solution which play an important part, particularly in Groves' gas battery, as is also shown in the works of Messrs. von Beetz, Gaugain and Morley.

Such is the total of the various causes which help to produce a secondary current in the platinum voltameter, in particular, and which apply equally to that with gold electrodes. They may possibly apply also to a slight extent, with regard to other metals; but we have seen that the oxidation of the positive electrode played the principal part in the voltameters previously considered.

32. **Voltameters with acidulated water, saturated with bichromate of potash.** — By using, instead of water acidulated with one tenth part sulphuric acid, a saturated solution of

bichromate of potash, equally acidulated, we notice phenomena which further demonstrate the influence of insoluble layers of matter deposited round the electrodes, or that of liquid sheaths, in increasing the resistance and diminishing the primary current in voltameters.

In this solution, silver and mercury become covered with deposits of insoluble red chromates which very nearly stop the passage of the current.

With the other metals, the reduction of the solution by hydrogen liberated around the negative pole, tends, no doubt, to increase the E.M.F. of the primary current; but, on the other hand, the reduced liquid forms, round this electrode, a sheath lower in conductivity, contributing to diminish it.

The secondary E.M.F. of these voltameters is a less important cause of falling off in the primary current than is the change in the solution. When the oxide of the metal is soluble, there is produced at the positive pole another saline sheath, equally liquid, which is only gradually dispersed and forms a further hindrance to the passage of the current.

The movement of either electrode also produces a very marked effect in these voltameters; the decomposed layers of solution are dispersed, the active solution finds its way more quickly to the electrodes and the current regains nearly all its original intensity.

This is, besides, what is noticed in Poggendorff's bichromate of potash battery, to which a fairly constant E.M.F. is given by the injection of air into the solution;—an improvement made by Messrs. Grenet and de Fonvielle.

By studying the phenomena which take place in voltameters we are able to account for those which are produced inside voltaic cells; because they are but the model, transposed, so to speak, into another apparatus where they may be examined under more favorable conditions.

33. **Conclusions.** — We may draw, from the above, the following conclusions :

The falling off in an electric current when a voltameter of acidulated water and electrodes of various metals is placed in the circuit, is due to several causes, acting with more or less intensity, according to the metals used. Sometimes these causes are all united.

1. The insolubility and bad conductibility of the oxide formed upon the positive pole.

2. If the oxide be soluble, a bed of saline liquid is produced by the decomposition and forms a sheath which prevents the the rest of the solution finding its way to the electrode.

3. The inverse, secondary, current (polarisation) which arises.

34. This secondary current, observed on closing the voltameter circuit, immediately the primary current is cut off, arises also from several causes :

1. With most metals, it is due, principally to the reduction of the oxide coating formed upon the positive electrode by the action of the primary current, and to the oxidation of the negative electrode which is brought to, or maintained in, a perfect metallic state by the liberation of hydrogen under the influence of the primary current.

2. With metals not easily oxidised, such as gold or platinum, the secondary current is due, for the most part, to the action which hydrogen has upon the negative pole, during the passage of the

primary current, whether by uniting itself with the metal of the electrode, or by modifying the chemical composition of the liquid with which it is surrounded, or by simply becoming dissolved in it to a small extent.

The compound thus formed, the solution thus changed, or the hydrogen gas dissolved, tend to combine again with the oxygen arising from decomposition of the water whilst the circuit of the voltameter was closed, and consequently furnish one of the elements of the E.M.F. of the secondary current.

But this secondary current is also due, at the same time, although in a less degree, to the slow oxidation in the positive electrode of the metals in question, during the passage of the primary current, on the one hand; and on the other, to the formation of strongly oxygenised compounds with the liquid in the voltameter; and finally, to a small proportion of oxygen gas dissolved in the water round the electrode. The slightly oxidised metal, the modified solution and the dissolved oxygen re-combine with the hydrogen when the secondary circuit is closed and furnish another element of the total E.M.F. of the voltameter.

35. The causes which contribute towards the developement of the secondary current in a voltameter are, as may be seen, very numerous. We could, strictly speaking, mention many more of a purely physical nature, resulting from the electrical condensation which must always take place in a system made up of two conducting materials of a certain nature, separated by one of a different nature. But, although the idea of a simple physical action may have arisen in the minds of physicists, in order to explain the production of a current without any apparent chemical action, and which may have given rise to the expression voltaic *polarisation*, this kind of

action may be entirely disregarded here, as much on account of the conductibility of the solution as the proximity of the surfaces in play, and the chemical actions just reviewed appear to us to be the principal causes of the production of the secondary current.

CHAPTER II.

Storage of the energy of the Voltaic Battery by means of Secondary Cells with lead plates.

―――

Various arrangements of these cells.—Their electro-chemical FORMATION *or preparation.—Their effects.—Preservation of their charge.—Residues.—Returns.*

―――

36. Since we have known that secondary currents constituted an important cause of the falling off in the E.M.F. of voltaic batteries, attention was devoted towards the prevention of these currents in the batteries themselves, and they were very fortunately neutralised in the first bi-liquid, and constant current batteries, due, as we know, to Becquerel.[1]

―――

(1) Annales de Chimie et de Physique, 2e série, t. XLI, p. 24, 1829.

Taking it from another point of view, we have sought to collect the secondary currents and to make use of them, thus accumulating the energy of the voltaic battery.

We have found, as may be seen in the preceding researches (20), that the secondary E.M.F. of a voltameter with lead plates in water acidulated with sulphuric acid, was higher and more lasting than that of any of the other metals, and that it surpassed that of the strongest voltaic element known—that of Grove or Bunsen.

With such an E.M.F., it only remained, in order to form a secondary element of great power, to endow it with a very low resistance, or to increase its surface to the greatest possible extent. This was so much the more easy from the fact that the two plates necessary to form it were of the same nature and of a metal so extremely flexible and malleable as lead.

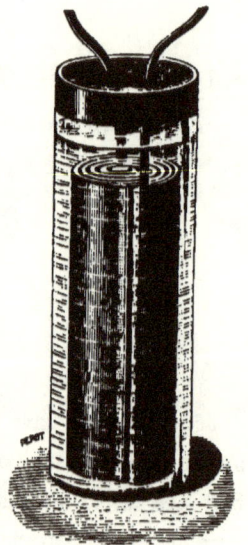

Fig. 7.

37. **Secondary cell with coiled lead plates.**—It is thus that we were led to construct, in 1860,[1] a secondary element of great power or quantity, by using an arrangement similar to that which Offershaus and Hare had employed for the voltaic cell proper, namely, by rolling two long, wide lead plates into a coil, separated one from the other by a thick cloth, and then immersing them in a glass jar full of water acidulated with a tenth part sulphuric acid.

Figure 7 shows the construction of a secondary cell of this kind.

---

(1) Comptes rendus, t. L, p. 640. Mars 1860.

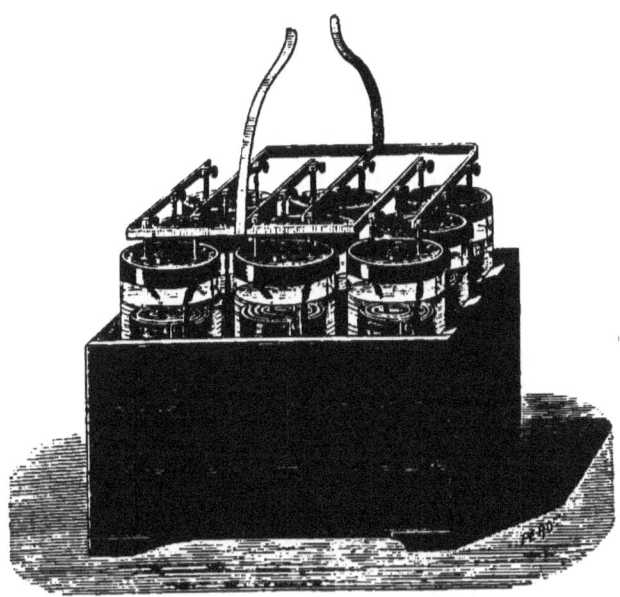

Fig. 8.

38. Secondary battery of large surface for quantity.
Figure 8 represents a secondary battery of nine cells, having a total surface of ten square metres, the earliest effects of which we exhibited at the Académie des Sciences, on the occasion of the meeting of March 26th, 1860.

By passing through this apparatus the current from five small Bunsen cells, we obtained, after a few minutes' action, a very bright spark possessed of great heating power when the two terminal wires of the battery in which the cells were coupled up, either in three parallels of three cells each in series, or all in parallel as shown in figure 8, were brought into contact for an instant.

39. Secondary cells with parallel lead plates.—But the arrangement of secondary cells in the form just described was somewhat objectionable by reason of the additional resistance introduced by the cloth intended to separate the electrodes. Moreover, this cloth becoming rotten in time, in the acidulated water, the lead plates might come into contact with each other, and so throw the cells out of order.

To overcome these objections, we used another arrangement[1] consisting of two series of parallel lead plates, with the terminals of the even row of plates joined on one side, and those of the odd row joined together on the other side, put into communication with the two poles of a primary battery.

These plates, brought very near to each other, and separated in the middle by insulating rods, were arranged vertically in a rectangular gutta percha cell, furnished with interior grooves, for holding the parallel lead plates.

In figure 9, which represents the plan of this arrangement, the letters $a\ b\ c$, and $a'\ b'\ c'$ show the two series of lead plates.

Fig. 9.

Figure 10 represents the apparatus complete. The terminals of the plates joined together at K and K' come through the upper part of the gutta percha cell, and may be connected with the primary battery by the wires P and P'. The cell having been filled with water acidulated with a tenth part of sulphuric acid, a cover is fused to the upper edges. A very little hole only, is left, in order to allow an escape for gases arising from electrolysis, during the passage of the primary current.

---

(1) Comptes rendus, t. LXVI, p. 1255.   Annales de Chimie et de Physique, 4th. series, t. XV, p. 10, 1868.

40. To illustrate the heating effects that can be produced with this apparatus, a thick platinum or iron wire f is fixed between the terminals B and O, of which B is connected with the plates $a\,b\,c$, by a small metal plate placed against one of the sides of the cell, not seen in the figure, and the other only put in contact with the second series of plates $a'\,b'\,c'$ through the intervention of a contact maker M. This contact maker is not intended to change the direction of the current, but simply to connect up the circuit, sometimes with the primary battery for charging, sometimes with the terminals O and B for discharging.[1]

Fig. 10.

41. By using six lead plates, twenty centimetres long by twenty two high, and considering that the double surfaces of all the plates are used, with the exception of the two outside surfaces, we have thus a small secondary battery for quantity, with a surface of about half a square metre, which will redden iron, steel, or platinum wires one millimetre in diameter, after being submitted for a short time to the action of a primary battery of two Bunsen cells.

42. The above arrangement, which we have used for several years, for obtaining at a moment's notice temporary currents of

---

(1) This arrangement of contact maker in the form of a simple metal bolt, rounded at the ends, which fits into small cast brass tubes, seemed a very convenient method, and we have used it since 1868, in several other ways, for making or breaking electrical communication at will.

considerable intensity, appeared to still possess some objections, which we have had to do our best to remedy.

The vertical gutta percha cells or jars became in time contracted, and caused the lead plates to approach each other very closely by buckling them, and causing contact.

This substance not being transparent prevented, besides, the sight of phenomena which take place inside the cells, and which it is important to be able to follow during the charge, as will be seen further on (60).

43. **Latest form of secondary cells with lead plates.**—We then returned to an arrangement very similar to the first described (37), but modifying the method of separating the lead plates.[1] We separated these plates no longer with a thick cloth, but by narrow strips of india rubber, presenting the advantage of not being injured in the acidulated water and only covering a very small portion of the surface of the electrodes.

Fig. 11.

Figure 11 represents the manner in which we proceeded to coil

---

(1) Comptes rendus, t. LXXIV, p. 592, 1872.

the plates close to each other without touching, and shows the arrangement of a couple thus constructed.[1]

Two pairs of india rubber strips about one centimetre wide by half a centimetre thick, are necessary to prevent the plates touching each other. The terminals are shaped at the opposite ends of the plates, in order to avoid as much as possible any contact, and to equalise the distribution of the primary current upon the surfaces of the electrodes, by separating the two points by which the positive and negative electricity flow into the secondary cell, as far as possible. But this is not indispensable if the plates are very uniformly rolled together. The chemical action of the primary current is then distributed equally over the whole surface of the secondary couple, even when the two poles are very close to each other.

We then coil the lead plates, thus separated by two or three pairs of india rubber strips, round a wood or metal cylinder, as shown in figure 11. As this cylinder is withdrawn as soon as the coil is finished, it does not matter of what it is made.

It is also better to place two small rubber strips at right angles to the others along the cylinder when beginning to roll the first turn, so as to well separate the edges of the two plates which might otherwise come into contact.

44. The coil once made, we take out the interior cylinder with care, and to make the whole more firm, the coils are permanently held in position by means of little gutta percha crosses softened by heat.

---

[1] We here enter into details, which will perhaps appear trifling, upon constructing and putting these cells to work; but the results achieved by their means in our more recent researches, have led us to wish scientists to make these cells easily for themselves, for the objects they might have in view. We are, besides, only acceding to a desire, often expressed, for the knowledge of all the details of the secondary cells.

The couple, thus constructed, is put into a cylindrical glass jar, and supported inside this jar by little gutta percha wedges. The jar is then filled with water acidulated with one tenth part sulphuric acid.

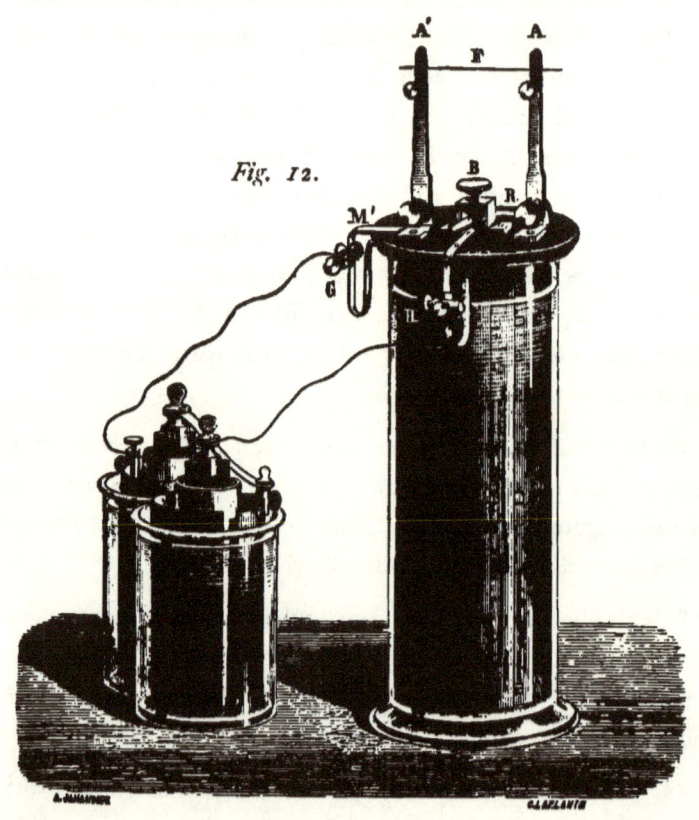

Fig. 12.

45.—Figure 12 represents a secondary cell of considerable size,[1] made according to our description, and also shows the arrangement of the connections for charging or discharging the cell, and illustrating the effects it can produce.[2]

---

[1] The lead plates are about sixty centimetres long, by twenty wide, and one millimetre thick.
[2] Les Mondes, t. XXVII, p. 426, and following, 1872.

The glass jar containing the lead plates immersed in acidulated water, is covered with a vulcanite disc which carries the metal parts intended to close the secondary circuit, when the cell is charged. The terminals of the two lead plates are connected by means of the binding screws G and H, at the same time, with a primary battery of two small Bunsen elements and with the little copper plates MM'.[1] The small plate M is fixed under another copper plate R, the extension of which, forming a spring, can be screwed down against the little plate M, which is then connected with the terminal A. The little plate M' on the other hand is in fixed communication with the terminal A', and between the arms of these two terminals are placed the wires intended to be reddened or fused by the secondary current. Wires from any other apparatus through which it is desired to pass the same current, may also be connected to these two terminals.

46. This system of connections, arranged in the simplest possible manner, so as to be held upon the cover of a jar, does not break the primary current when the secondary circuit is closed, like that which has been described above (40). If then we wish to try the effect of the secondary cell alone, it must be cut off from the primary battery by disconnecting the wires G and H, and then screwing up the button B. But if these wires are left in communication with the secondary cell when it is desired to discharge it by depressing the button, there is no

---

(1) The terminal tongues of the lead plates were previously covered whilst hot, throughout their length, with a thick varnish of spirits of turpentine, or with a composition of wax and resin, in order to prevent the acidulated liquid from creeping by capillary attraction, up to the connections with the copper plates. The terminals of the lead plates at this point, must be thoroughly brightened, and then covered with varnish or buried in a bed of composition poured upon a cork which seals the cell. This cork is pierced with a hole through which a little glass tube is passed.

objection; the apparatus works just as well; its effects are even slightly increased by the action of the primary battery, which is added to that of the secondary cell, when the circuit of the wire F is closed; for the two currents from the primary battery and the secondary cell, which are in opposition during the charge, become joined in parallel in the discharge.

47. We shall see further on (75), that the secondary cell, *formed* or prepared in a special manner, preserves its charge long enough to give strong effects, without the additional help from the current of the primary battery, during the short period of the discharge. But the continued connection of the secondary cell with the primary battery, presents, at least, the advantage that if we have a series of experiments to make with successive discharges of the secondary cell, all the intervals in which the discharge is not going on, become immediately utilised by the primary battery for charging the secondary cell.

48. **Chemical actions produced in secondary cells of lead plates.**—We have already studied (18) the chemical actions which take place in a voltameter with lead wires or plates, when the circuit of the voltameter is closed (short circuited) after the action of the primary current, and we have considered in detail the principal causes of the strong secondary current produced.

In the secondary cells which we have just described, these actions are naturally exhibited upon a larger scale, and produce phenomena, the study of which has been useful for endowing these cells with important qualities, such as furnishing discharges of long duration, of preserving their charge for a long time after the action of the primary battery, and of thus storing the chemical work of the voltaic pile.

49. When a secondary cell of large surface, such as that shown in Fig. 12, is new, that is, when the lead plates comprising it have never served to transmit the current in a voltameter, and it happens to have the current from two Bunsen cells passed through it, oxygen gas appears almost immediately upon the positive plate; some of it at the same time oxydises the surface of the plate, and this becomes quickly covered with a very thin coating of peroxide of lead.

On the other hand, the hydrogen, after having reduced the slight layer of oxide with which the lead was probably covered by exposure to the air, soon appears, and if, in a few moments, we try the secondary current given by the apparatus, we find it is already very strong by the sharpness of the spark produced when the secondary circuit is closed and opened again instantaneously, with a copper wire of low resistance. But the current thus obtained is of very short duration. It would certainly produce the incandescence of a very fine platinum wire, but would not redden a thick wire of the same metal. This happens from the layer of peroxide of lead on the surface of the positive plate being very thin, and as it becomes quickly reduced immediately the secondary circuit is closed, it cannot furnish a sufficient quantity of electricity.

50. But if, after having kept the circuit closed until the secondary current is exhausted, the apparatus be charged a second time, the plates are then in a condition little differing from that in which they were at the commencement. Whilst the secondary circuit is closed, the oxygen, being carried against the plate which

was negative during the passage of the primary current, has slightly peroxidised this plate, at the same time that the peroxide formed upon the other plate during the passage of the primary current, was reduced by the hydrogen. We have then, after a first experiment, two plates, the surfaces of which present a molecular condition differing from that in which they were when new. They are covered with thin layers of oxide and reduced metal respectively, which will facilitate the ulterior action of the primary current upon the secondary cell.

51. If we first consider the lead plate which was negative when the current passed through for the first time, it is as we have just seen, covered with a layer of oxide after the passage of the secondary current. The consequence is that if the primary current be passed through it again, the first portion of hydrogen will be devoted to reducing this layer of oxide, instead of the thinner layer resulting only from exposure to the air, as had happened previously. Then a period longer than the first will take place before the appearance of hydrogen upon the surface of this plate; for this gas will not begin to be set free till the oxide is completely reduced to a state of pulverulent or highly divided lead upon the surface of this plate.

52. If we study what takes place upon the positive plate for the second time, during the action of the primary current, the first portion of oxygen which tends to be set free at its surface encounters, this time, a layer of reduced peroxide or divided metallic lead, upon which this gas has more hold, so to speak, than upon the lead plate which serves for the first time; the gas is more easily absorbed, and we also begin to note a delay in the appearance of oxygen upon this plate, a period which corresponds

to the time necessary for re-oxidising the layer of reduced lead upon its surface.

When the secondary circuit is again made, the above described phenomena again take place, and it is easy to imagine that when these operations have been renewed a great number of times, the surfaces of lead in the secondary cell will be found in a more favorable condition for oxidation or reduction; the layers of oxide alternately formed or reduced, will become thicker, and the resulting secondary effects will show a longer duration and greater intensity. This is in fact what is observed: the more a secondary cell is charged and discharged, the greater is the duration of the secondary current.

53. Formation, or electro-chemical preparation, of secondary cells of lead plates.—We have thus attained the extention in duration of the discharge of secondary cells, by charging them successively a great number of times, and discharging them in proportion, so as to develope upon their surface, and even produce to a certain depth in the thickness of the plates, these layers of oxide and reduced metal, the finely divided state of which favors the development of the secondary current.

This result has also been obtained in a still more marked manner, by successively changing, several times, the direction of the primary current acting upon the secondary couple.

Everytime this change of direction is made, we notice, if there is a galvanometer in the circuit, a remarkable reinforcement of the primary current, for the first moments, which is easily explained because the E.M.F. of the secondary cell is added under these conditions, to that of the battery. But it is better, in order not to exhaust the primary battery too quickly, to short circuit the secondary cell before reversing the primary current; without that, the current from the battery is devoted for the first few moments to doing chemical work which could be effected by the discharge of the secondary cell itself.

Thus, the secondary cell must be first discharged, then recharged in the opposite direction. In this case, the primary current again acts directly upon surfaces which are nearly in the same electro-chemical condition; we do not then observe the increase of the current mentioned above, and there is no useless loss in the energy of the primary battery in the form of heat dispersed through the circuit.

54. We have seen moreover that it was advantageous, with a view to this preparation of the secondary cells, to allow a period of repose of several days between the reversals, in order to give to the deposits of oxide and reduced metal, time to attain a crystalline nature, and strong adherence to the surface of the plates. We observe in fact that the plates of the secondary couples which have undergone these actions acquire in time a peculiar crystalline appearance. The peroxidised plate, and the plate covered with reduced lead, are both of them strewn with shining points, and become changed in their molecular construction, not only at the surface, but some distance within the pores of the metal; we even notice that the peroxidised plate in particular ends by showing considerable fragility.

55. The plates thus changed lose none of their weight, by any number of charges and discharges. They only act, so to speak, as a base for the chemical actions which take place upon their surface, and which, constantly following each other in opposite

directions, cannot consume them. The lead is continually oxidised and reduced at the same time that the water is alternately decomposed and vice versâ.

56. These combined operations that we have called by the name of *formation* of secondary couples, and which consist, as we have just seen, in *forming* or maturing them, in order to produce deposits of a certain thickness, allow of our obtaining heating effects of considerable duration in the discharge.

Taking a secondary cell with lead plates of half a square metre in surface, *formed* or matured to a desired extent, and previously charged by two Bunsen elements for three quarters of an hour, we can redden for eight or ten minutes a platinum wire one millimetre in diameter, and seven or eight centimetres long, and a wire of about half a millimetre in diameter, for twenty to twenty five minutes.

57. Prolonged immersion of the lead plates in the acidulated water, before the action of the primary current, hastens materially the formation of secondary cells.

This fact seems to be explained by the slow penetration of the liquid into the interior of the pores of the metal, which allows the electrolytic action to be exercised deeper, and to produce a greater quantity of oxide or reduced metal.

58. The intensity of the primary current also has a great influence over the more or less perfect *formation* of secondary cells. The current produced by two Bunsen elements is what experience has proved to be the most suitable. A too feeble current only produces very superficial deposits, and the nature of the peroxide

of lead produced upon the positive plate is different as to its physical aspect (and perhaps in a chemical point of view),[1] from that of the peroxide formed by a stronger current. The resulting oxide from a sufficiently prolonged feeble current is black; that which a stronger current creates has a clear brown color characterising the true peroxide of lead.

Daniell elements, even in a large number, do not *form* secondary cells so well as two Grove or Bunsen elements possessing less total E.M.F., but giving a larger quantity of electricity. The oxidations and reductions must be made at a certain rate, and the electrolysis must be sufficiently vigorous to thoroughly penetrate the interior of the metal.

59. To form the secondary cells well, it is expedient to bear in mind the preceding remarks, and to take into consideration above all that the action of *time* is indispensable, as in many chemical actions of which nature and industry present examples.

It is a kind of electro-chemical tanning, if one may use such an expression, which the electrodes of secondary batteries undergo. The lead plates should be penetrated little by little, as deeply as possible by the oxidising and reducing actions of the primary current, so as to completely change their molecular formation.

---

[1] We have had occasion to notice in voltameters with copper electrodes, an example of the difference in the chemical nature of the oxide formed, according to the E.M.F. of the current used. Thus with a current from two Bunsen elements, there is obtained at the positive pole, black bioxide of copper, which only appears for an instant at the commencement of the electrolysis (8) and is then dissolved in the liquid without being visible, according to the degree of its formation; whilst if fifteen elements are used a spray of reddish oxide resembling protoxide of copper is formed at the extremity of the positive electrode, and falls to the bottom of the voltameter without being immediately dissolved in the liquid.
(Bibl. univ. de Genève, t. VII p. 332, 1860.)

The periods of rest mentioned above (54) between the changes in direction of the primary current have the greatest effect. Thus a secondary cell the plates of which have been submitted for several hours at once to the action of the primary current, being left to itself for a month, without being discharged, then taken in hand again and charged in the reverse direction, will give a discharge of double the duration which it could give before. The following, in fact, is the plan to be pursued in these operations;

The secondary cell being filled, to begin with, with water acidulated by a tenth part of pure sulphuric acid,[1] we allow the current from two Bunsen elements to go through it six or eight times, in different directions alternately, on the first day. The secondary cell is discharged between each change of direction, and it is easy to take note, either by the incandesence of a platinum wire or by any other plan, that the length of discharge continues regularly to increase.

Gradually we increase the time during which the secondary cell is submitted to the action of the primary current in the same direction. This period is successively extended, after the first day, from a quarter of an hour, to half an hour and an hour. We leave it finally charged in one direction until the next day. The following day it is charged for two hours in the reverse direction, then in the first direction, and so on. We still notice an increase in the time of the discharge, but a point soon arrives beyond which this duration does not perceptibly increase, especially when the primary battery, not having been renewed, becomes gradually

---

[1] It is essential that the sulphuric acid contain no traces of nitric acid, which would attack the lead and help to cause the terminal tongues of the plates of secondary batteries to be broken off.

weakened by these successive operations, and has no longer sufficient intensity for the electrolysis to penetrate deeper into the interior of the plates (58).

We then leave the secondary cell at rest for eight days, and it is then recharged in the opposite direction for several hours, without making for that day any fresh reversal. Gradually the period of rest is extended to a fortnight, a month, two months, &c., and the duration of the discharge continues to increase. There is no limit to it but the thickness of the lead plates.

The positive plate, if it is thin, ends by being almost completely transformed, in time, into peroxide of lead of a crystalline nature. The negative plate becomes gradually changed, to a certain depth beneath its surface, to reduced lead, spongy and crystalline.

It is not, of course, necessary to push the electro-chemical preparation of the secondary cells to this complete transformation of the physical and chemical nature of the plates; for the plates would then acquire a higher resistance, and require more time for charging.

When the secondary cells give a current of sufficient duration for the purposes to which they are to be applied, there is no further occasion to change the direction of the primary current each time they are used. The store of peroxide of lead accumulated on the positive plate would take too long to reduce, and no effect would be obtained from the cell for several hours. One direction is then adopted in which the secondary cells are always charged, once they are sufficiently *formed*.

60. **Absorption of the gases during the charge of secondary cells.** — When secondary cells prepared under the

foregoing conditions are charged, we observe that the gases are completely absorbed, for a certain time, to such a point, that, with a secondary couple having a surface of one square metre, submitted to the action of two Bunsen elements, it takes twenty to thirty minutes before any gas appears upon the surface of the plates.

The entire work of the primary current is stored in the apparatus in the form of oxidation of lead on the one hand, and reduction of the lead oxidised by the previous discharge of the secondary current, on the other hand, and may be again given back (excepting an unavoidable loss) in the form of secondary current, by the inverse reconstitution of the same products. When the gases begin to be liberated, it is a sign that the battery is no longer doing useful work towards producing the secondary current. Thus the sign of the gases being set free in a secondary cell, previously well *formed*, becomes an indication of the maximum charge that the cell can take, and there is no further advantage gained by prolonging the action of the primary current, as far as regards the secondary effects. The cell must be in this condition of previous electro-chemical preparation, in order that the liberation of gas may indicate that it is fully charged; for, with a new secondary couple, submitted for the first time to the action of a primary current, or with a couple already *formed*, but which has remained a long time without being used, we see the gas set free at the surface of the plates, almost immediately, without having attained the maximum degree of charge which they are able to take.[1]

---

[1] New lead plates exhibit, however, to a certain extent, these signs of the absorbtion of gas, after the direction of the primary current has been changed two or three times only. They pass successively through the lightest and darkest shades of peroxide of lead, or attain metallic tints of a silver grey, according to the gas which acts upon their surface. The gases are absorbed for the short time corresponding to the development of these thin layers of oxide and reduced metal.

61. **Maintenance of secondary cells.**—When a secondary cell is considered sufficiently *formed*, intervals of rest of several months, far from being useful, as in the operation of *formation*, would tend to increase the resistance of the cells, and to render their charge longer and more difficult. It is consequently preferable to charge them from time to time, or to keep them constantly charged by a weak current, so as to avoid the production upon the surface of the positive plate, of a layer of protoxide of lead, of low conductivity, arising from the slow and spontaneous production of the peroxide of lead.

62. Secondary elements thus *formed* by the current from two Bunsen cells, may be charged and kept in order by means of a battery of three Daniell or Callaud elements filled with pure water, such as those used in telegraphy. The charge obtained, it is true, is not so strong as with nitric acid elements, but the use of this primary source is more convenient in a great number of cases.

63. **Effects produced by secondary cells with lead plates.** — Temporary calorific and magnetic effects may be obtained, with the secondary cells that we have just described, far more intense than the primary battery used to charge them could produce. We shall show their application in the second part of this work. The apparatus shown, (fig. 12) (45) has been arranged specially with a view to demonstrate calorific effects, and we have already mentioned one or two of these effects (40-41-56). By coupling up four or five of these cells in parallel, thick wires of iron or steel may be melted, and molten globules may be obtained 7 to 8 millimetres in diameter. In order to form the globules, the binding screws of the apparatus must be drawn together gradually as the wire begins to melt.

64. We have made a special study of these globules, on account of some remarkable analogies which they seem to show, and which we shall point out in the fourth part. By examining them with the help of a magnifying glass, darkened, whilst they are in a state of fusion and throwing out a brilliant lustre, under the influence of the powerful electric current passing

Fig. 13.

Fig. 14.

they be, still present a small body of matter, brought to a very high temperature, in which the calorific movement, itself arising from the previous electrical motion, is not immediately destroyed; the chemical results of the rise in temperature (such as the oxidation of the carbon, if the wire is of steel) follow also for an instant the passage of the electric current, and are shown by the gaseous bubbles, seen upon the surface of the globules.

No doubt these effects could be produced by any other source of dynamic electricity, in sufficient quantity; but we describe them here as being obtained with secondary cells, in an easier and more convenient manner for study, than by any other means.

65. **Magnetic effects.**—The magnetic effects produced by discharging secondary cells are also very powerful. Electro magnets wound with thick wire may be more strongly excited in this way than by ordinary batteries in which there is much less surface, and which, consequently, could not afford, in a given time, so great a quantity of electricity. Powerful permanent magnets may be easily formed,—so much the more, as, in this case, the current only needs to be passed for an instant through the coil surrounding the steel bars intended to be magnetised.

Electro-dynamic experiments, in which the conducting wires have to carry the greatest possible quantity of electricity, may be successfully repeated with the aid of secondary cells.

If we want to work an electro-magnetic motor, not for a work of long continuance, but to carry out some experiments, or for illustration, these cells may be employed with great advantage. When the circuit of the electro-magnets is not too low in resistance, the current of the secondary cell is not so quickly spent, and we have been able, even with secondary cells of much smaller surface than that represented in figure 12, in a single discharge, to drive an electro-magnetic motor for more than an an hour.

These cells may be also used for working induction coils, and carrying out a great number of experiments, by means of a single discharge.

66. Formation of ozone in secondary cells, and voltameters with lead plates.—We have noticed, when using secondary cells with lead plates, that they often gave off a strong odour of ozone, especially when they were charged in the reverse direction by means of a rather strong current.

Nonoxidisable metals, such as gold and platinum, used to be considered as the only ones which could be used for obtaining ozone by the electroysis of water. By studying voltameters with lead electrodes, from this point of view, we found that ozone could be as easily produced with lead electrodes as with platinum, and even in a stronger degree.[1]

This may easily be proved by taking two voltameters, one of which is made with platinum wire, the other of lead, of the same length and diameter, and sending through them both, the current from ten Bunsen elements. By immersing some strips of paper iodised and starched, in the open tubes placed above the positive wire in each voltameter, we see them both grow blue, and it may be observed that the paper immersed in the oxygen of the lead wire voltameter grows blue more rapidly, and with more intensity than the paper immersed in the oxygen of the platinum wire voltameter.

The pungent smell, and the rapidity of the oxidation of silver, also show an easily noticeable difference. By causing ozoned

---

(1) Comptes rendus, t. LXIII, p. 181, 1866.

oxygen to be liberated simultaneously from two voltameters of similar solutions of iodide of potassium, the solution submitted to the action of the oxygen of the lead wire voltameter turns yellow more rapidly than that which is acted upon by the oxygen of the platinum wire voltameter, and we find the quantity of iodine liberated by the ozone from the platinum wire voltameter is only about two thirds as much as is given by the ozone from the lead wire voltameter.

The lead wire voltameter should be slightly *formed* beforehand, like the secondary couples (53), so that the layer of peroxide of lead, produced at the positive pole, thoroughly covers the metal.

67. This appearance of ozone, more abundant with lead electrodes than with those of platinum, is a rather difficult fact to explain in the present state of our knowledge of ozone.

Still we think it may be explained by taking into consideration the influence which the greater or less metallic condition of the electrode must possess over the generation of ozone. We know with what readiness metals, or oxidisable materials in general, absorb ozone; how silver which is not tarnished by the atmosphere, blackens under its influence.

We may then presume that platinum itself is not absolutely free from action upon ozone, in proportion as it is generated in a voltameter with wires of this metal, and that a certain part of the ozone generated can be destroyed, whilst, in a voltameter with lead electrodes, the positive, once well coated with insoluble peroxide, in acidulated water, forms a less metallic electrode than platinum, and is not so favorable to the destruction of ozone.

This appears to be proved by the fact, that in comparing the quantity of ozone produced by wires and plates of lead, we have obtained more ozone with wires than with plates, and, by using at the positive pole of a voltameter, a very short and very fine point of lead, we have had more ozone than with a thicker wire of a moderate length. We may then take it, that the less the surface presented by the electrode, the greater is the quantity of ozone produced, so that, if it were possible to decompose water without an electrode, or with as non-metallic or non-oxidisable electrode as possible, we should have the maximum proportion of ozone in a voltameter. The formation of ozone by the simple influence of static electricity, as described by Messrs. Frémy and Ed. Becquerel, or by the flow through the induction coil, in the tubes of Messrs. Siemens, Houzeau, de Babo, Bollot, Thénard, Berthelot, etc., would seem to support this theory.

68. Shades or tints of oxide produced at the positive [negative] pole during the discharge of secondary cells.

—We have remarked above, when studying the voltameter with lead electrodes (18), that the reduction of the peroxide of lead, formed upon the positive electrode, under the action of the primary current, was not the only chemical action produced, nor the sole source of the E.M.F. of the secondary current; but that the other lead electrode became oxidised at the same time, in consequence of the decomposition of the water, in the interior of the secondary cell itself.

This oxidation, hardly visible in a voltameter, is clearly shown in secondary cells, by a very distinct phenomenon, during the discharge.

If, for example, we discharge one of these cells, by reddening a platinum wire, the negative plate at first preserves, in the exterior visible part, the clear grey tint of metallic lead, during nearly the whole time that the incandescence lasts; but, upon the wire ceasing to be red, a dark shade is seen to appear which covers the exterior surface of the plate and gives it a darker grey tint. The oxidation of this plate, by the interior current of the secondary cell, is not sufficiently complete, nor sufficiently prolonged to give it the shade of peroxide of lead; but its change in physical aspect is nevertheless appreciable, and reveals the chemical phase produced. During the greater part of the discharge, the oxidation takes place inside the coil; towards the end it extends gradually over the entire plate and, naturally, the outside which has not the other electrode opposite to it, is the last to be affected.

69. We also notice, especially in the earlier periods of their *formation*, the liberation of gas, which is shewn in certain voltameters (13) upon breaking the primary current. In this case the phenomenon is still more marked, especially if we close the secondary circuit with a short thick wire of good conductivity, on account of the intensity of the current which is generated in the interior of the cell.

70. **Duration of the discharge in secondary cells.**—The length of the discharge of secondary cells is in proportion to the degree of *formation* which they have attained. Thus, one of these couples will maintain a platinum wire, one millimetre in diameter, at a red heat for from 1 to 10 minutes, according to its degree of *formation*. But, taking the same couple, the duration of the discharge of course depends also upon the resistance of the exterior circuit. With a secondary cell which would give, by using a thick platinum wire, only an incandescence of a few minutes, we can keep a platinum wire ·5 of a millimetre in diameter, incandescent for an hour.

The length of the discharge of secondary cells consequently depends upon the extent of their surface, the thickness of the deposits on the plates—especially the layer of peroxide of lead, which penetrates the positive plate—and, lastly, the resistance of the outer circuit.

71. **Constancy in the secondary current during the discharge.**[1] — The resistance opposed to the passage of the secondary current, in which the quantity of electricity has more importance than the tension, plays the part of moderator or regulator, and thus transforms the effect of a current of a temporary nature, into a relatively constant current of a long duration. Thus with a resistance of 50 metres of copper wire 1 millimetre in diameter, placed in the circuit of a secondary couple and a tangent galvanometer, the coil of which had a resistance equal to that of three metres, we obtained a practically constant intensity of current for about an hour.

72. This constancy on the part of a current, which one would suppose on first thought must continually decrease from the time the circuit is closed, is explained by the great quantity of electricity stored, under the form of chemical work, in the secondary cell, with a relatively low E.M.F., just as a very broad tank, containing a great quantity of liquid of very little depth, would give, for a long time, a nearly constant outflow through a small orifice, and cease quickly when the liquid fell to the level of the orifice, so, a secondary couple of large surface, whether it reddens a wire or is discharged through a galvanometer, only shows a falling off in intensity some moments before completely ceasing to supply a current.

---

(1) Les Mondes, t. XXVII, p. 474. 1872.

The fact is startling when we discharge the secondary cell through a fine platinum wire. The incandescence is maintained for a long time at a uniform degree, and ceases almost abruptly when the store of chemical work accumulated in the cell is exhausted.

73. We have proved this in a still clearer manner by tracing the intensity curve of the secondary current, during the length of the discharge effected through a constant resistance. The curve presents, for the greater portion of its length, nearly a straight line, parallel to the base upon which the time is marked, only falling abruptly towards the end of the discharge.

74. We except here the effect produced during the first moments of the discharge following the break of the primary current. Immediately after this break, there is always a maximum effect, owing to the double origin from which the secondary E.M.F. arises. This force, as we know from studying the voltameters, (28, 31, 34,) is due to chemical actions exercised by the primary current upon the electrodes and the liquid surrounding them at the same time. The products resulting from this latter action, such as oxygenised water, formed in a very small quantity, very uncertain, and not very adherent to the electrodes, are immediately reduced, or dispersed in the remainder of the liquid. Then, even when the secondary circuit is left open, their action disappears, as we shall see further on, when treating of the E.M.F. of secondary couples.

On the other hand, the products resulting from the primary current on the electrodes, are formed in a certain quantity, remain fixed to the electrodes, and only alter when the secondary circuit is closed.

Hence, two actions help to produce the secondary current in the cells in question : one only acting during the first moments following the break of the primary current, the other being prolonged for an hour.

It is this latter effect—this continuous current from the secondary cells—which manifests the constancy noticed above.

75. Preservation of the charge taken by secondary cells.—Lead plate secondary cells acquire a valuable property by the work of *formation:* that of preserving a great portion of their charge for a considerable time after the action of the primary current.

Thus, a secondary pair, well *formed* and thoroughly charged, will raise a platinum wire, a millimetre in diameter, to a red heat for several minutes, two or three weeks after having been charged. We have even obtained this effect sometimes, with cells exceptionally well *formed*, more than a month after having subjected them to the action of the primary current.

76. If the peroxide of lead deposited on the positive plate did not tend to reduce itself spontaneously in the acidulated water, by local action with the metal beneath it, the preservation of the charge held by the secondary cell ought to be unlimited. But this peroxide is reduced, and with so much the more facility as the layer is thinner, and in a new secondary couple the charge[1] cannot be preserved at all.

---

(1) We employ here the word *charge*, for want of a more exact term, in order to designate the effect resulting from the accumulation of the chemical work by the primary battery in the secondary couple. No doubt there may also be a veritable static charge, as in a condenser, but quite unimportant, in spite of the large surfaces, on account of the conductivity of the electrolyte, which separates the two metallic plates (35).

How is it then preserved, with very little loss, in a *formed* secondary cell?

It may be explained by taking into consideration that, if the deposit of peroxide of lead has a certain thickness, the surface in immediate contact with the solution is alone reduced to a state of protoxide, and then protects the underlying layers from any action.

77. It seems difficult, at first sight, to imagine a non-conducting coat of oxide preserving an electrode from chemical action when there is no current passing through the apparatus, and powerless to protect it when there is an electric circuit formed. But we know, on the other hand, how powerful is the intervention of electric energy to modify, or determine, actions which could not take place without it, and we will quote, *en passant*, an instance in which an analogous effect may be easily set forth.

If we employ, for example, a voltameter with acidulated water in which the positive pole is formed of a bed of mercury, and the negative pole by a platinum wire, there is produced at the surface of the mercury, from the first moments of the passage of the principal current, a layer of insoluble sulphate of mercury which soon weakens, in a marked manner, the intensity of the current. If we reverse the current, this superficial coating no longer offers any opposition to the electro-chemical action; the sulphate of mercury is immediately cleared from the surface of the mercury as if it were swept by a draught of air; it becomes soluble in the liquid, whilst the surface of the mercury becomes brighter and soon gives off a steady liberation of hydrogen.

In a lead plate secondary cell the operations take place in the same manner, when the primary current is sufficiently strong and when it has just been reversed. The deposits are sometimes loosened, and fall in flakes to the bottom of the liquid.

We then conclude that, when the secondary circuit is closed, the electrical actions called into play cause chemical re-actions, such as reduction or oxidation, beneath the non-conducting coatings which act as a protection when an electric circuit is not made.

78. **Residual charge afforded by secondary cells.**—Secondary couples, when discharged, are able to give at the end of a certain time, without having been again charged, a residual charge, similar to that obtained from Leyden jars.

If for example we raise a platinum wire to a red heat by the discharge of one of these cells, and if the circuit be broken immediately the wire has ceased to be red, we can, in a quarter or half an hour, again observe an incandescence of several seconds, by closing the secondary circuit.

If the cell has been very well charged, and if the discharge has lasted a long time, we can obtain, the next day and several following days, a residual discharge, the duration of which may be from two to three minutes.

One could even still obtain a further series of successive discharges, decreasing in intensity. This phenomenon arises from the fact of the layer of peroxide of lead upon the positive plate not being reduced throughout its thickness by the first discharge of the secondary cell.

In fact, during this first discharge, whilst the peroxidised plate is being reduced, the other plate becomes oxidised, as we have described (50), and tends to produce an opposing current in the interior of the pair itself. The E.M.F. of this current, which may be distinguished by the name *tertiary*, ends by equalling the E.M.F. of the pair; the two plates are soon found in an electro-chemical state, nearly identical, and the discharge of the couple appears terminated. But if the circuit be again broken, the *thin* layer of oxide formed during the discharge, upon the lead plate previously negative, is reduced little by little in the acidulated water, as we have also explained above (76). At the end of a certain time (more or less long according to the duration of the discharge of the secondary cell), this layer is completely or partially reduced, and, as on the other hand, the *thick* coat of peroxide of lead developed upon the positive plate by the primary battery has not been entirely reduced throughout its depth during the discharge, the two plates are found to be again in a different electro-chemical condition, and consequently, a fresh discharge may be obtained by closing the circuit. The successive residual discharges, decreasing in intensity, afterwards obtained, may be explained in the same way.

79. **Increase of the intensity of a secondary cell, with rest, after being charged.**—A still more remarkable phenomenon, sometimes presented by secondary cells, is the following :—

We take a cell which has been left a long time without being recharged. As has been already described (61), these cells do not recharge easily. We send through it for several hours the current from two Bunsen elements, and when we find that the cell will heat a fine platinum wire, without being able to redden it, the primary circuit is broken. At the end of twenty hours rest, it is noticed that the cell raises this same platinum wire to a red heat,

Thus, the cell appears to attain, by repose an E.M.F. higher than that which the primary battery was able to give it.

We think that this fact may be accounted for as follows:— during the time that the primary current passes through the secondary cell, the electrolytic gases tend to set themselves free between the metal and the layers of oxide,—more or less reduced,— which cover it. These gases not being able to escape easily, increase by their presence the resistance of the secondary couple, by preventing contact of the liquid with the surface of the plates, in proportion as they are oxidised or reduced by the primary current. If we suspend the action of this current the gases are slowly liberated, the metallic surfaces changed by the primary current are brought into better contact with the liquid, and the secondary cell is able to give, in this quite exceptional case, a discharge of a greater intensity after the repose than immediately after the action of the primary current.

80. **The Electro-motive force of lead plate secondary cells.**—The E.M.F. of lead plate secondary cells is somewhat easier to measure than that of a voltameter, because the duration of the discharge is longer, on account of the greater surface of the couples, and the larger quantity of deposits accumulated. In all cases, there is, as we have seen (74), a maximum of E.M.F. immediately following the break in the primary current, which is of rather short duration on account of the peculiar nature of the chemical actions which produce it. It is then necessary, for measuring this maximum, to operate in nearly the same way as with voltameters (6), that is to say, to close the secondary circuit as quickly as possible after breaking the primary circuit, and to examine the first effect produced upon the galvanometer.

We have carried out this measure in various ways, either by means of a tangent galvanometer and changeable resistance, or with the electro-magnetic scale (9), in operating upon a single secondary couple or upon a large number connected in series.

In one experiment among others, made upon 40 secondary elements, all charged at once, as will be seen further on (ch. III), by three Bunsen cells, coupled up in series, at the moment of discharge, we have obtained an attraction of the electro-magnetic scale equal to 9·45 grammes, which corresponds to 0·236-gr. per secondary cell. The E.M.F. of one Bunsen element, measured by means of the same scale, was found equal to 0·164-gr.

If this E.M.F. be taken as a unit, that of the lead plate secondary cell will be, by deduction, 1·44. In operating upon a single well formed secondary cell, there is necessarily a more perfect charge, even with two Bunsen elements as primary source instead of three, as in the preceding experiment, and most of the figures we have obtained are comprised between 1·45 and 1·50.

We may, then, consider the electro-motive force of lead plate secondary cells, observed immediately after breaking the primary current, as about equal to one and a half times that of a Bunsen element, and about two and a half times that of a Daniell element.

It is, besides, the result which we obtained with a simple voltameter (20).

81. If we measure this E.M.F. two or three minutes after breaking the primary current, then, even when the secondary circuit has remained open, it is found to be noticeably diminished, and reduced to 1·17, by reason of the causes which produce a temporary polarisation, as previously mentioned (74), disappearing.

But the E.M.F. is maintained very constant at this point during nearly the whole of the discharge.

82. This difference between the initial E.M.F. of a lead plate secondary cell, and its subsequent E.M.F., is, besides, easily appreciated when a secondary couple is discharged through a platinum wire immediately after the action of the primary battery. For the first moment the incandescence is very high, almost fusing the wire. If we allow, on the other hand, an interval of a few minutes between the action of the primary current and the discharge, the incandescence is not so high but very uniform until the end of the discharge of the secondary cell (73).

83. **Resistance of lead plate secondary cells.**—We first found out the resistance of the secondary cells[1] by a plan similar to that sometimes used for measuring the resistance of voltameters, namely, opposing two secondary cells to each other, by means of a special commutator at the instant the primary current is broken, so that their E.M.F. may be neutralised and the double resistance alone put into play.

But since we have attained, by the means of the *formation*, a discharge of some duration and certain constancy, we have been able to measure the resistance of these cells, previously well charged, like ordinary constant current cells.

The method of employing a tangent galvanometer and variable metallic resistances, not having afforded us very consistent figures, we have latterly given preference to a plan, based upon the use of derived currents, due to Sir William Thomson, simplified by M. Mouton,[2] and suggested, with some reason, as one of the most

---

(1) Annales de Chimie et de Physique, 4e série, t. XV, p. 19, 1868.

(2) Journal de Physique, par ch. d'Almeida, t. V, p. 144, 1876.

convenient and rapid that can be employed, for measuring the resistance of voltaic cells.

We have found that the resistance of secondary cells of the various dimensions which we have used, varied from 2 to 5 metres of a copper wire 1 millimetre in diameter.

We have noticed that the extent of surface, or the size of the secondary couple, which has a great effect upon the duration of the discharge, has far less influence over the resistance of the cells, than the distance the plates are apart, their more or less perfect degree of *formation*, and their condition. Thus, cells of very small surface (2 square decimetres) in which the platesw ere only separated by a distance of 2 millimetres, had but a resistance of about 3 metres of copper wire 1 millimetre in diameter, whilst cells having half a square metre of surface, the plates of which were 5 or 6 millimetres apart, and which had remained a long time without use, possessed a resistance of 4 to 5 metres of the same wire.

However, in any case, these experiments prove that the resistance of the secondary cells is very low, and the intensity of effects obtained from them may be thus explained.

84. **Necessary E.M.F. of the primary current.**—We may draw from the preceding tests relative to the E.M.F. (80), that, to charge lead plate secondary cells, it is only necessary to employ a primary current of a higher E.M.F. than one and a half times that of a Bunsen element. And two of these elements are perfectly suitable for the purpose, as we have already seen (41). Three elements would, no doubt, charge the secondary cell more quickly; but this excess of E.M.F. is not necessary and may even

prove inconvenient, if the cells have been *formed* with a weaker current, in loosening the coatings of oxide and reduced metal upon the plates by a too rapid libération of gas.

If it be desired to employ Daniell elements as the source of the primary current, three of them are sufficient to exceed the opposing E.M.F. of the secondary cell; but, as the excess is not so great as in the case of two Bunsen elements, the secondary cell is not so completely charged. Besides, they afford, in a given time, a less quantity of electricity. Also, the charging occupies very much longer time, and as there may be causes for loss, of which we will speak further on (92), these elements do not charge the secondary cells so completely. They do very well, all the same, for keeping them charged (62).

85. **Limit of the charge that secondary cells may take.**—It would seem that, by increasing indefinitely the extent of the surface in a cell, or by coupling a number of cells together, one might obtain, with a weak primary battery, a secondary current of an indefinite intensity. But there is a limit, beyond which we cannot go, in prolonging the duration of the charge. Just as we cannot thoroughly charge batteries of condensers (in static electricity), of large surface, by means of very small electrical machines, on account of the losses through the atmosphere when the charge takes too long, so, in this case, there exists a cause of loss whilst the charging goes on, in the tendency of the peroxide of lead to be spontaneously reduced in the body of the acidulated water, whilst it is being formed (76).

This reduction is so much the easier when the layer is more slowly deposited, and consequently thinner, so that a time comes when the action of the primary current, in order to renew or maintain this layer upon the surface of the plate, is balanced by the tendency of the peroxide of lead to be reduced in the liquid. The limit of the charge that the secondary cell can take from a certain primary source is thus fixed.

86. **Secondary cell charged by a thermo-pile.**—Any apparatus giving a continuous current of electricity in the same direction may be used to charge secondary cells, provided it has sufficient E.M.F. as we have just seen (84). Thus, for example, a secondary cell may be charged by one of Ed. Becquerel's or Clamond's thermo-piles, and if there be produced, in discharging it, the incandescence of a platinum wire, a portion of the actual heat used for the charge, which is stored in the form of electro-chemical work in the secondary cell, is expended in this manner.

87. **Secondary cell charged and discharged by means of the Gramme machine.**—A secondary cell may be equally charged by means of mechanical work, and the charge may be reconverted into the same form, any length of time after having been carried out.

This is the result of an experiment, which we have made together with M. Alfred Niaudet,[1] with the help of the Gramme machine, which gives, as we know, a continuous current in the same direction; and being reversible, like magneto-electric machines in general, can serve as an electro-magnetic motor.

Figure 15 represents this experiment. A Gramme machine with Jamin magnet is coupled up to a secondary cell, which may be, for the purpose of illustration, of much smaller dimensions than that described above (45).

---

[1] An electro-dynamic experiment by Messrs. G. Planté and A. Niaudet. Comptes-rendus, t. LXXVI, p. 1259, 1873.

If, after having charged the cell, the machine be stopped, without breaking the connection between the two parts of the apparatus, we see it immediately start in motion under the influence of the discharge, thus proving a fact which seems at first sight contradictory : namely, that the machine turns, not in the opposite direction, but in the same direction it was running when charging the secondary cell.

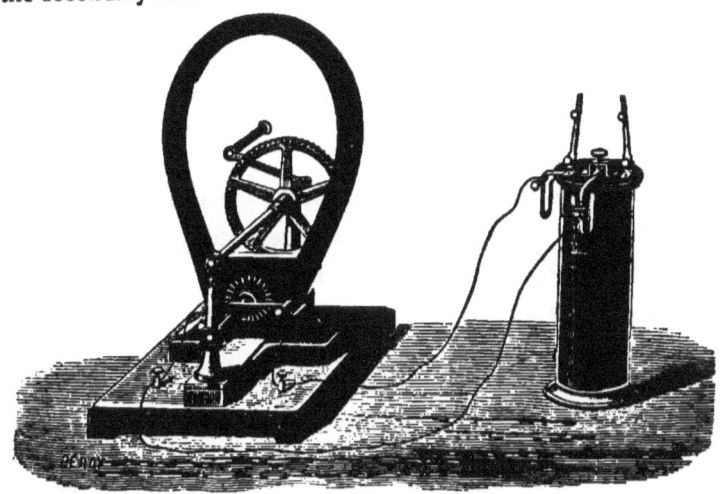

Fig. 15.

This fact is explained in the following manner :—if we first consider the direction of the current furnished by the machine, then that of the current given back by the secondary cell (which is in the opposite direction to the preceding one), and if we take into account the actions resulting from it, we see that according to the laws of induction and electro-dynamics, the motion of rotation ought certainly to be in the direction shown by the experiment.* On the other hand, it is necessary to mention, that in charging the secondary cell, the Gramme machine, driven at a high speed,

---

* We presume it is a dynamo-electric machine and not one with a permanent magnet which the Author refers to in this case. *(Note by Translator.)*

attains a higher E.M.F. than the opposing force developed in the cell. When the machine is stopped, and the cell, by discharging, causes it to revolve in the same direction, the speed given to it is not high enough to cause it to generate a higher E.M.F. than that possessed by the cell itself.

The rotation, then, takes place by reason of a difference between the two opposing electro-motive forces: that of the charged secondary cell, which predominates, and the weaker one which the machine tends to develope by its movement, under the influence of the discharge of the secondary cell.

We may say that, in this experiment, the Gramme machine, which produces the same effect as a battery immediately it is set in motion, polarises under the influence of the discharge of the secondary cell, since it presents an opposing E.M.F., so that we thus find a mechanical reproduction, so to speak, of voltaic polarisation.

88. **Various analogies presented by secondary cells.**—On taking into consideration the effects produced by secondary cells, of which we have just exhibited the principal properties, we find that these apparatus can play the same part in dynamic electricity, as the Leyden jar and condensers in static electricity. The analogy continues, even, as we have seen, to the residual discharges which they give.

Nevertheless, we have remarked (48) that the cause of the production of the current in the secondary cells was purely chemical; that if there were condensation of electricity, properly called, this effect was not worthy of notice, and that these apparatus did not directly store the electricity itself but the chemical work of the battery. We will not then insist any more upon this analogy, but will point out a few of another order, offering no less degree of interest.

89. These secondary cells may be also used in connection [compared] with all apparatus employed in mechanics, for the accumulation of work resulting from the action of force, such as hydraulic accumulators, compressed air reservoirs, springs (so called secondary motors), pulley blocks, winches, &c., going back to the simplest appliance, the lever itself.

A secondary couple is, in fact, a kind of lever for dynamic electricity; because one can thereby obtain with a weak electrical power, an increase of this power, in such proportion as one wishes, on the condition of loss in speed, or a necessary sacrifice of time, in order to accumulate the effect.

The same principles as those which apply to the lever ought to be taken into consideration; otherwise, the secondary cells might cause both illusions and labours, the uselessness of which could be demonstrated in the same way as the impossibility of perpetual motion.

90. One of the principal advantages presented by these secondary cells is to afford storage for spare electrical work, or as one sometimes expresses it, at the present time, a powerful [potential] energy that may be used at will, in a longer or shorter time.

We have considered, up to the present, that this power is expended in a shorter time than was required for its accumulation, so as to obtain an effect of greater intensity than that of the primary force. But it may also be interesting, in certain cases, to charge a secondary cell with a greater force, in a very short time, and to expend the accumulated work during a more extended period.

The following experiment is, from this point of view, very illustrative.

We take as a primary electric source a battery of two Bunsen elements, strong enough to raise a platinum wire, half a millimetre in diameter, to a red heat, and we charge by this means, for one minute only, a well *formed* secondary cell. We then discharge the secondary cell through a considerably finer wire than that reddened by the primary current, a wire ·1 millimetre, for example, and we notice that this wire continues red for about five minutes.

The expenditure of the stored work has, in this case, lasted much longer than was required for its accumulation. An effect is thus realised analogous to that produced by sharply drawing the cord wound round a child's top, and then letting the toy expend, by a prolonged rotary motion, the motive force which was communicated to it in one brief instant.

91. Return obtained from Secondary cells.—Taking this view of secondary cells, by comparing them with apparatus for accumulating mechanical work, leads us to measure their efficiency; or the proportion which the electrical work restored by their discharge bears to the electrical work expended in charging them.

The work most directly effected by the voltaic current being chemical, we have compared the total of the chemical actions produced in the circuit during the charge, with that of the same kind of actions produced in the discharge. As a means of comparison we chose the reduction of sulphate of copper, it being the easiest electro-chemical action to measure.

A sulphate of copper cell, furnished with a platinum plate, previously weighed, was joined to the primary battery of two Bunsen elements; on the other side, a well *formed* secondary cell was placed in connection with the primary battery for a certain time, and the passage of the primary current was stopped immediately the liberation of gas began to appear in the secondary cell, the latter then being, as we have seen above (60), charged nearly to saturation. (The platinum plate of the test cell, coated with copper, was weighed after the experiment.)

This done, we discharged the secondary cell by closing its circuit through a voltameter of sulphate of copper, provided with another platinum plate, already weighed, and we did not stop the experiment until the action of the secondary current was completely exhausted. We considered this result reached when the deviation of a galvanometer, also in the circuit, was reduced to zero.

By comparing, according to the deposits of copper obtained, the total chemical work returned by the secondary cell during its discharge, with the total chemical work expended in charging it, we found that the proportion of return was from 88 to 89 per cent.

92. The loss of work corresponding to the 11 or 12 per cent missing in the return may be accounted for as follows:

First, the spontaneous reduction in the acidulated water of a small portion of peroxide of lead, in proportion as it is deposited upon the positive plate; a cause so much the more important as the surface of the secondary cell is greater, and the deposited layer consequently thinner, and as the charge lasts longer. With a cell of extremely large surface in proportion to that of the primary battery intended to charge it, this cause of loss would increase to an indefinite extent, and it would be hardly possible to charge the cell.

Secondly, the incomplete *formation* of the secondary cell: a portion of the gases being then driven off without producing any useful chemical effect in the ulterior development of the secondary current.

Thirdly, the polarisation or the development of opposing E.M.F. in the interior of the secondary cell itself, whilst it is at work. The consequence is, that when discharging the cell, we lose the portion of work wasted from this cause. In order to estimate this loss, it would be necessary to ascertain the work which could be produced by the residual charge (78), and add it to the return.

However that may be, in spite of these sources of loss, we see, according to the return obtained—nearly equal to 90 per cent.—that a lead plate secondary cell, well *formed*, affords a very perfect accumulator of the work of the voltaic battery.

CHAPTER III.

Transformation of the energy of the Voltaic Battery by means of lead plate Secondary Batteries.

Secondary Batteries of high tension.—Various arrangements.—Their effects.—Instructions regarding the use of Secondary Batteries.—Analogies.

---

93. The results we have just shown allow, as we have seen, of the accumulation of a *quantity* of electricity arising from a given voltaic source, without obtaining a higher *tension* or E.M.F. than that of the source. But many electrical demonstrations require not only a quantity of electricity, but a considerable tension. It was, then, interesting, to try to obtain, in an easy manner, and without too much loss in the transformation, some effects of a higher E.M.F. than that of a given electric source.

Grove's gas battery offered the first means of approaching the solution of the problem. In fact, by charging successively a certain number of these gas cells, by means of the same primary battery, so as to fill them with the gases arising from electrolysis, a battery is formed of a higher E.M.F. than that of the primary one. But the gas battery being able to give only a very small quantity of electricity, and each of the cells having, besides, but a very low E.M.F., this solution of the question presented more interest from a theoretical than from a practical point of view; because it was difficult to take advantage of, even for scientific research.

The apparatus known by the name of the De la Rive electrochemical condenser,[1] permits of the production of the electrolysis of water in a voltameter with platinum electrodes, by employing the extra current developed in an induction coil with a single voltaic cell; a result which could not be attained directly from the cell itself, and proving in consequence, the development of an E.M.F. higher than that of this cell.

The works of Poggendorf[2] show that we can obtain, by means of several voltameters with platinum electrodes, polarised by a given current, an increase still more marked, and even indefinite, in the E.M.F. of the current, by collecting, with the help of a pivoted mercurial commutator, the polarisation current emanating successively from all the voltameters, connected in parallel and in series.

---

(1) Archives de l'Electricité, t. III, p. 159, 1843; et Traité d'Electricité par A. De la Rive, t. I, p. 391.

(2) Annales de Poggendorf, t. LX, p. 568, 1843; et t. LXI, p. 586, 1844.

M. J. Müller[1] employed, for the same object, a spring commutator, without mercury, which permitted of a continuous rotary motion.

M. Thomson's polarisation battery[2] also offered a solution of the problem, by allowing a series of platinum voltameters to be successively charged with great rapidity, and discharged with the same rapidity, so as to finally obtain a continuous current of a higher E.M.F. than that of the primary battery used to charge them.

94. We have in our turn applied[3] lead plate secondary cells, as described in the preceding chapter, to the production of a current of higher E.M.F. than that of the primary battery, by using their opposing E.M.F., already somewhat raised in itself, and the persistency of its action after the passage of the primary current.

Lead offering, besides, the advantage of being easily arranged with large surface, we have been able to produce far higher effects, both in E.M.F. and quantity simultaneously, than those of the primary current, thus transforming and accumulating the work of the voltaic battery at one and the same time.

95. **Secondary battery for tension effects, formed of parallel lead plates.**—Fig. 16 shows the first arrangement we used. Secondary pairs to the number of 40[4] were formed, each

---

(1) Traité de Galvanisme et d'Electro-magnétisme, by G. Wiedemann, 2nd edition, t. I, p. 657.

(2) Annales de Poggendorf, t. CXXIV, p. 163, 1865.

(3) Annales de Chimie et de physique, 4th série, t. XV, p. 22, 1868.

(4) Only 20 cells are represented, in order to give distinctness to the figure.

composed of two lead plates, o,2oc. × o,2oc. in size, held in very narrow gutta percha cells and immersed in acidulated water. Each of the lead plates terminated with little copper plates which had springs at the two extremities and could be pressed, either by the metal bars MM', NN', or by an insulated bar BB' furnished underneath with metal pieces. These bars were arranged so as to form a frame to which an oscillating motion could be given.

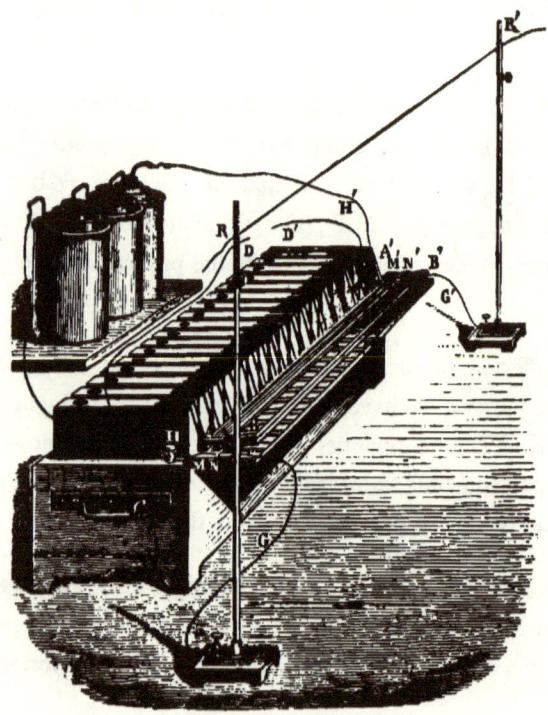

Fig. 16.

In the position in which the frame is represented by Fig. 16, all the springs connected with the positive plates are pressed by the rod MM', and all those which communicate with the negative plates are pressed by the rod NN'. The secondary couples are thus connected in parallel, or for quantity, and are charged through the medium of the wires HH', from a battery of three Bunsen elements placed near the apparatus.

When the insulated bar BB' is lowered, the metal portions of its lower surface thus press and unite, two and two, the springs communicating with the adjacent poles of opposite names of all the secondary couples. The battery is then in series.

By connecting the two terminal springs by means of the wires GG', to the metal rods provided with nippers, a platinum wire two metres long and ·4 of a millimetre in diameter, can be maintained at a red heat for from one to two minutes.[1]

96. Secondary battery for currents of high tension, formed of couples composed of coiled lead plates.

—As the cells of the battery we have just described presented, in time, some of those faults above pointed out (42), we replaced them by pairs of plates rolled into a spiral form, constructed as seen in Fig. 11, and have arranged the apparatus as represented in Fig. 17.[2]

Twenty secondary couples, held in cylindrical glass jars filled with acidulated water, are placed in two rows, and are connected to the springs of a commutator similar to the preceding one, intended to alternately unite the cells in parallel during the charge, and in series for the discharge.

Two copper cylinders, CC, C'C' are fixed to a bar of insulating material (wood or vulcanite) furnished with small metallic plates,

---

[1] When the secondary couples were new, a platinum wire of ·2 of a millimetre in diameter only could be rendered incandescent, and only for some seconds; but as the couples are formed by use, it has been possible to incandesce, for a longer time, wires of double that diameter.

[2] Les Mondes, vol: 27, p. 427, 1872.

so that they may be turned simultaneously in either direction by

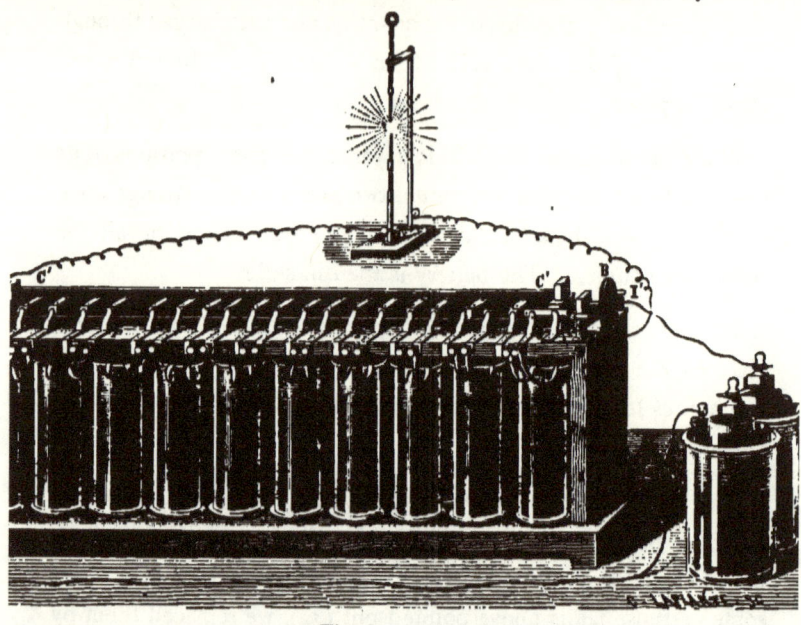

Fig. 17.

means of a button B, and rub, alternately with the bar, against the springs r r r.[1]

The combination of all the secondary cells in parallel during

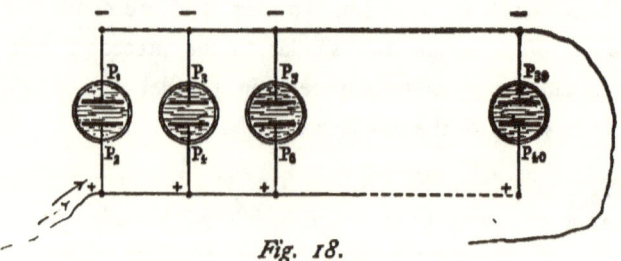

Fig. 18.

the charge, is represented by the diagram Fig. 18, in which, to

---

[1] This commutator has been cleverly made by M. J. Morin.

simplify it, the cells are shown by two plates. The plates of the odd row P_1, P_3, P_5 and P_{39} are connected all together, and the plates of the even row P_2, P_4, P_6 P_{40} on the other hand, are also connected together, when the springs rub against the metallic cylinders, here represented by two simple lines, to which are attached the wires from the primary battery.

The plates of the even row thus become all positive, during the charge, and those of the uneven row negative. The combination of the secondary couples in series for the discharge is represented by Fig. 19.

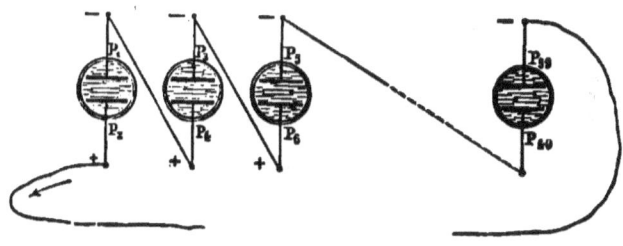

Fig. 19.

When the springs rub against the metallic parts of the insulated bar, all the couples are connected by their poles of opposite names and the discharge can be taken from the two end plates, the direction being opposite to that of the primary current, as indicated by the arrows in the two figures.

Fig. 17 (see p. 78) represents the battery producing a voltaic arc, the commutator being turned into the position which it occupies during the discharge.

To recharge the battery, the commutator must be turned a quarter of a revolution. All the cells, then connected so as to form but one of very large surface, are submitted to the action of the primary current, supplied by two Bunsen elements, the poles of which are connected with the terminals I and I'.

97. The coiled plates of the cells in the battery are 12 centimetres broad and 18 long. The space between the plates is from 3 to 4 millimetres and their useful surface is about 8 square decimetres. The resistance of each of these couples when charged is equal to 8 metres, 77 centimetres of copper wire, 1 millimetre in diameter, or about equal to that of a Bunsen cell of the same surface.

But, as we have said previously (83), this resistance may vary considerably according to the degree of formation of the plates, and the distance which separates them. Some couples of even smaller dimensions may present less resistance.

98. Figure 20 represents a secondary battery of smaller size than the last, and of a more simple construction, which we finally adopted for investigating the effects of electric currents of high tension.

The commutator is simply a wooden bar, on the edges of which are placed bands of copper, and metal pins pierce the wood at right angles. There is some similarity between this arrangement and the commutator of Cooke & Wheatstone's needle telegraph.

In the position of the bar shown by Fig. 21, the longitudinal copper bands gg', seen in section, touch all the springs such as rr', and unite all the couples in parallel; the metallic pins, one of which is represented by hh', and which traverse the bar, are insulated from the circuit. In the other position of the bar (fig. 22), the pins as shown at hh', touch the same springs rr', and unite all the couples in series. The surface of the small couples of this battery has been reduced to about two square decimetres, so as to take less time in charging them. The lead plates of which they are composed are in fact only 10 by 6 centimetres.

Fig. 20.

But as they are very near together, the resistance to the conductivity of these couples is very small. The commutator is represented, fig. 20, in the position it must occupy during the charge of the battery.

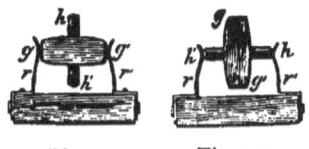

Fig. 21.   *Fig. 22.*

The binding screws QQ' communicate with the longitudinal copper bands of the bar and serve to redden or fuse short thick wires when the secondary couples connected in parallel have been submitted for a short time to the action of the primary current.

Similar binding screws may be adapted to the commutator cylinder of the battery already described: (96), fig. 17. The binding screws TT', fig. 20, join the outer poles of the battery, and allow the long fine wires to become incandescent or melted, when the commutator is turned so as to unite all the couples in series.

99. **Effects produced by secondary batteries composed of lead plates.**—We have just quoted, in describing each battery, some of the effects which can be produced; we will further add, that the duration of these effects depends on the more or less complete formation of the secondary couples which compose them (53). The potential of the current which they can supply of course depends on the number of couples. The E.M.F. of each cell being equal, as has been seen (80), to about one and a half Bunsen elements, and as, on the other hand, the resistance, with equal surfaces, is practically the same as that of the Bunsen element (97), there may be obtained, with a battery of forty couples, or with two batteries of twenty couples united, the same effect during the first instant of the discharge, as with a Bunsen battery of about sixty elements.

Experiments which take but a short time may often be repeated with a single charge. We may mention, among others, the melting of a steel wire, which can be effected, in a length of 1 metre 20 centimetres, with batteries of 40 couples, or $\overset{?}{x}$ m. 6.°cent. with 20 couples.

It is noticed that this fusion is accompanied by the formation of a chaplet of little molten metallic globules, visible when looking at the wire through darkened glass, or when examining fragments of the wire, broken after fusion. Fig. 23 represents this effect produced with the little secondary battery of twenty couples, last described, and we shall have occasion to refer to this experiment (IVth part) in order to account for the appearance presented sometimes by certain natural phenomena.

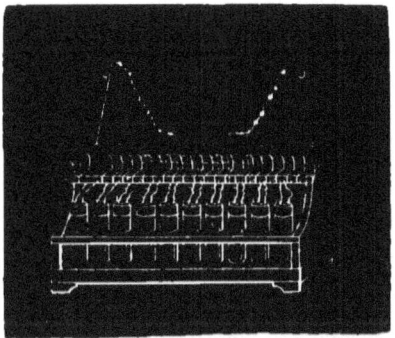

Fig. 23.

Without speaking here of effects which we describe further on and which are obtained by a large number of batteries united, we will just mention, among experiments which may be repeated with secondary batteries, that of light produced by the vaporization of mercury.

By placing a metal cup containing a few drops of mercury, in connection with one of the poles of a battery of twenty or forty couples, under a cover through which passes a rod terminated by a platinum wire, there is obtained, by bringing this wire in contact with the mercury, a beautiful light, which may be prolonged for three minutes.

Electro-chemic reactions which require a current of considerable tension, such as the decomposition of potash or sodium, may be demonstrated by means of these batteries.

100. **Large secondary battery of extended surface for effects of quantity or tension.**—In taking secondary couples

of large surface, such as that represented in fig. 12 (45), each being half a square metre in area, instead of cells of 2 to 8 square decimetres, and by arranging them as above, effects of quantity and tension are obtained simultaneously which last a considerable time, and thus is realised the simultaneous accumulation and transformation of the work of the voltaic cell.

With six large elements, well formed and arranged in the large surface battery represented in fig. 8 (38), but with a commutator placed above, so as to be able to unite them in series, after the charge in parallel, a voltaic arc is obtained which lasts from seven to eight minutes, endowed with greater brilliancy than that from a battery of equal E.M.F. formed of Bunsen elements of ordinary dimensions.

101. **Secondary battery of lead plates for prolonged effects of tension.** — We have also arranged[1] a secondary battery of lead plates, composed of 40 elements of very small surface (a few square centimetres each), for obtaining continued effects of tension, for which we will point out a purpose farther on (123).

In this case, there is only a simple transformation of the work of the primary battery, without accumulation. Tension is produced at the expense of quantity of electricity, and it is then better that the two Bunsen elements destined to charge the battery should be of rather large size.

We will limit ourselves to mentioning this apparatus, described in the memoirs quoted, intending to modify the arrangement in order to simplify the application.

---

(1) Annales de Chimie et de Physique, 4th series, vol. XY, p. 27, 1868.

102. **Instructions respecting the use of secondary batteries.** — To obtain the maximum effect that the secondary batteries previously described can give, it is well to be sure that each couple is in good condition, that is to say, sufficiently 'formed,' that there is no interior contact between the plates, and that the terminal tongues which serve as poles are not broken.

The effect produced by each couple of the battery previously charged may be tried separately and quickly by first turning the commutator a little, so that the springs joining the poles of all the couples are not in contact with any metallic part, and then by touching the poles with the two small copper plates of a little rheoscope (fig. 24) formed of a platinum wire stretched between two separate nippers. If this wire is from ·2 to ·3 of a millimetre in diameter, and 4 to 5 cent. in length, each couple of the battery ought to redden it brightly for about a minute.

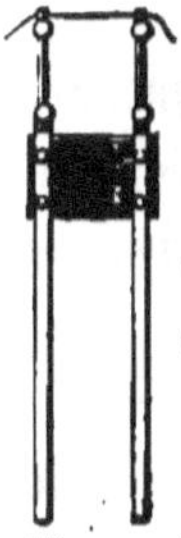

If a couple should not redden it at all, this couple ought to be examined to find out if there is an interior contact or one of its poles broken.

Fig. 24.

103. The case of interior contact or short circuit rarely happens if the couples are well made. It is besides easily recognised, for contact in a single couple would prevent all the battery from charging, by presenting much less resistance to the primary current than the other couples where the plates are separated by acidulated water. If, then, other couples of the battery are found charged, it is certain that the couple suspected has no interior contact.

It may be further proved by touching, for a few moments, the springs in connection with this couple with the two extremities of the wires from the primary battery. If there is an escape of gas, the couple has evidently no contact. If there is no escape of gas, there may be contact in the couple, or it may be broken.

In this case, the spark on breaking the circuit of the primary battery should be examined, the secondary couple being in the circuit. If this spark equals in vividness that which the battery alone would produce, there must be an interior contact. If there is no spark, it indicates, coinciding with the absence of escape of gas, that there is a pole broken.[1]

104. The fracture of a pole is the accident which is most likely to happen to secondary cells of lead plates.

The metallic porosity which plays so large a part in the operation of the formation of couples, by allowing the electro-chemic action to work to a certain depth, has, on the other hand, the disadvantage of allowing acidulated water to penetrate to the interior of the small terminal plates of the cells, and to gradually advance, until it comes in contact with the wires or small copper plates which bind the couple to the springs of the battery. Copper is attacked especially when in presence of another metal, and the connections are broken or impaired.

The creeping of the solution is further facilitated by the phenomenon of "electric carrying"[2] during the discharge, of the same kind as that observed by MM. Reuss, Porret, Becquerel and G. Wiedemann, which works from the positive to the negative pole; and it may here be remarked that the accident we speak of happens nearly always to the positive terminals of the secondary couples.

---

(1) However, if the battery is weak and the couples have remained a long time without working, so as to offer, at first, rather a high resistance, it must not be hastily concluded that the couple has a pole broken; one must wait some time after the current has passed, and try if a secondary current is produced by means of the rheoscope of platinum wire.

(2) Known as "Porret's' phenomenon." (*Translator.*)

These small positive plates are also liable to break at the level of the acidulated water, doubtless on account of local action between the immersed portion of the peroxydised lead plate and the exterior portion less peroxydised. In short, after the discharge of the secondary couples in a circuit of very low resistance, these terminal tongues may become powerfully heated, and thus acquire extreme fragility. In experiments we have made with numerous batteries, of which we shall speak further on (III. part), this heating of the terminal tongues of those couples which are not so well charged as others, and submitted to the action of all the other couples, may reach incandescence and fusing of the plates with a kind of explosion arising from the rapid vaporization of part of the liquid in the cell.

105. This accident may be prevented by using tolerably thick lead (1 millimetre in thickness), and by varnishing the tongues of lead, whilst hot, considerably below the point where they emerge from the liquid (45).

When this accident happens, it may be remedied by opening and washing the secondary cell, and by shaping another tongue out of a portion of the lead plate which was immersed in the liquid. The tongue thus cut, must be bent round, varnished, scraped at the end, and fastened by a little binding screw or nut, either varnished or enveloped in a bed of cement, on to the wire meant to connect the secondary couple to the terminals of the battery.

106. If it be desired to throw a couple, to which an accident has happened, out of circuit, without detaching it from the apparatus, it is only necessary to unite the screws of the springs which are

next each other by a bridge formed of a very small copper plate with the ends rasped like forks. In this way the circuit presents no interruption when the apparatus is discharged in series.

107. When there are found, among the couples of a battery, some which become less charged than others, in consequence of inevitable variations in the resistance of these different couples, they are thus known:—During the discharge, instead of contributing to the development of the secondary current, they act, relatively, as simple voltameters, and only give an abundance of gas under the influence of the secondary current produced by the other cells. These couples become equal with time, but if they are wanted to be charged separately and to be more formed, so as to put them on a level with the others, it suffices, without detaching them from the apparatus, to place a band of paper between the springs of the other couples, and the corresponding metallic portions of the commutator, so as to allow only those couples to be charged. These couples, if too numerous, would hinder demonstration of the effects of the battery, for they would absorb a part of the force of the secondary current during the discharge; it is necessary then for the battery to be equalized and this comes with time if the battery be frequently worked.

108. **Returns.—Comparisons.** — The secondary batteries above described, being more especially arranged to produce effects of tension, afford a return inferior to that of batteries intended for effects of quantity, and less capable of accurate measurement because of the differences of resistance of the various couples of which they are composed. These are not however, the less efficacious as instruments of transformation, which, after the action of a weak current, permit of obtaining for a considerable time, the most intense effects of the voltaic battery.

They may be compared with several machines used in mechanics for transforming and accumulating force, particularly the machine known by the name of the "pile driver." Indeed, in this latter machine, a heavy weight, raised gradually to a great height by a series of successive efforts, is let drop, and performs by its fall, in one instantaneous blow, the principal part of the work done during a considerable time.

In the secondary batteries of which we are speaking, the amount of chemical action produced by a weak current of electricity, distributed over a great number of couples, developes an amount of electro motive force, which, united upon the closing of the secondary circuit, gives back, in the form of a very intense current of short duration, the amount of work accumulated during the time that the charge from the battery lasted. The effects of quantity correspond with the fall of a very heavy mass, only raised to a slight height; effects of tension are analogous to the fall of a mass less heavy raised to a great height.

These comparisons also show the connection which exists between divers manifestations of force, or of motion in general, and the variety of effects which may be obtained, by analogy, with electric force.[1]

---

[1] On the employment of secondary currents for accumulating or transforming effects of the voltaic pile. Comptes rendus, vol. LXXIV, p. 592, 1872.

SECOND PART.

APPLICATIONS.

Uses in Galvanocaustics and Therapeutics in General;—Exploding Mines;—Domestic Uses;— Electric Breaks;—Signalling by means of Lights; etc.

109. Galvanocaustic applications.—The calorific effects of the secondary cells, above described, may be utilised in galvanocaustics, for operations which do not require an action of long duration, and such cases often occur in therapeutics. We pointed this out in 1868[1] and put it into practice in 1872, after finding that it was possible to increase the charge of secondary cells by the process of "formation," and also to endow them with the property of preserving the charge for a considerable time without great loss.

---

(1) Researches in secondary currents and their application.—Annales de Chimie et de Physique, 4e série, t. XV, p. 21.

Fig. 25 represents the arrangement of a secondary cell for this purpose. The cell is enclosed in a box, the top of which is fitted with metal plates connected to the two poles where the conductors of the cauterising apparatus terminate. The essential portion of the latter is a platinum wire doubled into a point or twisted, according to the object required to be effected.

Secondary cells arranged in this way, once charged, may be easily carried about, and will afford without any handling in the presence of the patient, the heat necessary for the operation.

If the operations take only a short time, the store of electrical energy in the secondary cell suffices for carrying out several of them without recharging.

Cauterizations of the lacrymal gland have been effected by Dr. Ominus, in 1873, upon seven or eight subjects successively, without it having been necessary to recharge the apparatus.

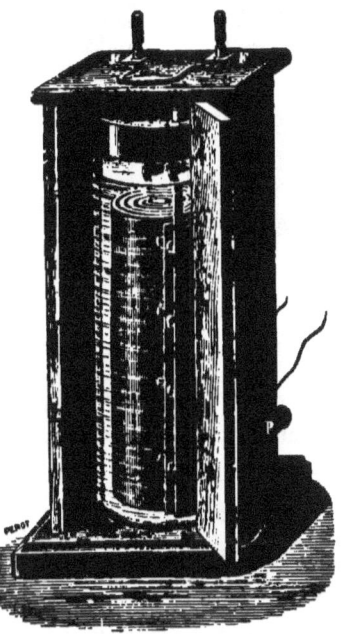

Fig. 25.

When one of these cells, of the dimensions above described (45), has been well "formed," it will raise to incandescence a platinum wire 1 m.m. in diameter by 7 or 8 centimetres long, for eight or ten minutes, and a wire of the same length and half a millimetre in diameter for more than twenty minutes.

110. For small operations, such as occur in dental surgery, a smaller cell may be used, still more easily carried, such as that shown in fig. 26.

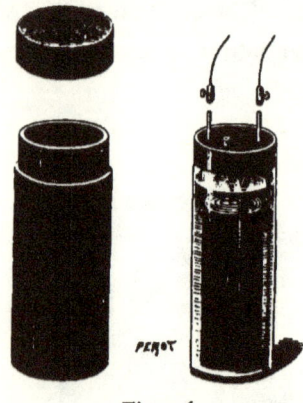

Fig. 26.

These cells are of the same dimensions as those composing the battery shown in fig. 17 (96 & 97). They may be completely sealed by means of a small plug of india rubber fitted in the glass tube which passes through the stopper of the cell, shut up in a case and easily carried in the pocket.

These cells, when well formed, will redden a platinum wire half a millimetre thick for two or three minutes, and a wire ·2 m.m. thick for five or six minutes.

Dr. Moret used these little cells, with success, in the treatment of neuralgia by means of cauterisation, of the kind known as *trans-current* cauterization, and for instantaneously stopping arterial hemorrhage.[1]

111. This latter type of secondary cell (fig. 26), with only a comparitively small surface, may be kept sufficiently charged by means of three Daniell elements. For cells of larger size (fig. 25), this source would be too weak and we recommend two Bunsen cells with the zinc in pure water, instead of acidulated, and the carbon surrounded by nitric acid in a porous jar. Such elements are, of course, weaker than those made up in the ordinary manner; the

---

(1) Revue de Thérapeutique, 44e année, p. 172. Avril 1877.

Among the physicians who were the earliest to interest themselves in this application and who made use of these secondary cells in their operations, we are happy to mention the names of Drs. de Bonnéfoux, Lailler, Lubinoff, Morétin, Constantin Paul, de Tavel, etc

zinc is only attacked slowly and feebly by the acid which finds its way through the porous jars; but the trouble of amalgamation is avoided, and these elements will keep in condition for charging secondary cells for about a week without it being necessary to dismount and renew them every day. More time only (about four or five hours), is needed, when charging with these cells, than when Bunsen elements of amalgamated zinc in acidulated water are used.

It is essential to always charge secondary cells in the same direction, as we have already stated, that is to say, to connect the positive pole of the primary battery with the peroxidised lead plate, and the negative pole with the lead plate preserved in a pure metallic state or coated with reduced pulverulent lead.

It is also of importance to keep the secondary cells open during the charging and only to entirely close them when carried about; for the gases given off during electrolysis would gradually cause a considerable pressure upon the liquid if the cells were closed, and would tend to make it creep along the lead terminals or exude by the smallest cracks in the stopper and thus affect the points of contact with the copper strips to which are fitted the cauterizer conductors.

Fig. 27.

Omission of this precaution is a frequent cause of deterioration in the connections of secondary cells.

112. **Use in lighting dark cavities in the human body and obscure hollows in general.**—If the discharge from a secondary cell be made to pass through a platinum wire doubled at the middle into a point, as shewn in fig. 27, the wire, being less quickly cooled, in this form, by the surrounding air, attains a degree close upon fusion of the metal, and it emits, by its incandescence, a very bright light which may be made use of.

In various conferences upon the effects created by these secondary cells, held in 1872 and during the following years, we had an opportunity of lighting the assembly for from half an hour to an hour with two secondary cells thus arranged, each of which gave out, with a platinum wire ·2 m.m. in diameter, a light nearly equal to that of a candle and with a very constant intensity (71).

M. Trouvé recently applied secondary cells for laryngoscopy, to light the obscure cavities of the human body and obscure cavities in general.[1] He arranged platinum wires of different forms, in the focus of little spherical, concave or parabolical reflectors, ingeniously combined according to the nature of the cavity he wished to light.

In order to prevent fusion of the platinum wires by too bright an incandescence under the action of the current from the secondary cell, M. Trouvé has added to the apparatus which we have before described, fig. 25 (109), a platinum wire rhéostat intended to graduate the intensity of the current, according to the diameter and the length of the platinum wire used as lighting apparatus, or cauterizer in galvanocaustics.

---

[1] Bulletin of the meeting of the Société française de Physique p. 2, Jan. 4th, 1878.— La Nature, 6th year, p. 107, July 13th, 1878.

He has also added a double circuit galvanometer for reading the charge of the secondary couple, and to discover in what state the battery for charging it happens to be.

113. Application to firing mines, etc. — Among the applications to which the couples of secondary batteries may be put is the firing of mines, for it requires a calorific effect of short duration repeated at certain intervals.

We described in 1868[1] a secondary battery of small surface which could be employed for this purpose, when the circuit has considerable resistance. The battery represented in fig. 20, may suit still better It is only necessary, in making use of these batteries, to keep them charged, and not to leave them too long without working, for they are then more difficult to charge.

A fuse formed by a platinum wire of $\frac{1}{10}$ of a millimetre, dipped in powder or gun cotton, may be fired by the current from a battery of 20 couples, with a resistance in the circuit equivalent to about 6 kilomètres of telegraph wire.

If the circuit has a lower resistance, say 300 metres of this same conductor, one secondary couple may suffice to fire successively, with a single discharge, a considerable number of fuses. It is not necessary for these cells to have a large surface. Couples such as those represented in fig. 26 (110), or even yet smaller, such as those in fig. 31 (115), may be used and constitute a very portable apparatus.

114. Fig. 28 represents a portable battery enclosing two small secondary couples, which connect easily in parallel during the

---

(1) Annales de Chimie et de Physique, 4th series, vol. XV, p. 26.

charge, and in tension during the discharge, by means of a combination of three buttons CDC which serve as a simplified commutator in this particular case.

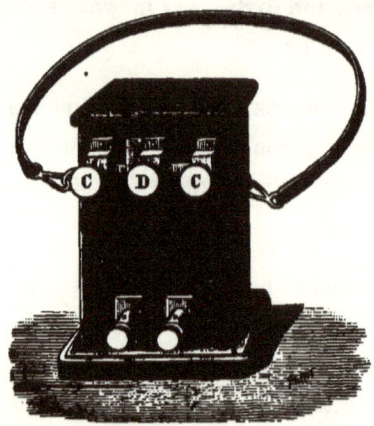

Fig. 28.

The connections of the secondary couple are, in fact, arranged so that, the poles of the primary battery being put in connection with the two edges of the box, if the two buttons C C are pressed, the two couples become charged at the same time, like a single couple of double surface, and if the button D be pressed after having loosed the buttons C C, and the connections with the primary battery taken away, the two secondary couples become combined in tension for the discharge. These buttons are otherwise arranged like that represented in fig. 12 (45).

If it is wished to charge each couple separately, independently of each other, only one button of three should be pressed, either the left button for charging the couple on the right, or the right button for charging the couple which is behind the left button.

The three buttons ought never to be pressed all at once, in that case the apparatus would be short circuited.

These small batteries of two cells may be used in cases where, the circuit being of too high resistance, ignition would not be sufficiently instantaneous with a single secondary cell.

When simultaneous firing of a great number of fuses is required, it is merely necessary to multiply the number of the cells or secondary batteries, without increasing the power of the primary battery which serves to charge them, and which is always composed either of two Bunsen elements mounted in the usual way, or, as we have previously described, three Daniell elements.

It is true that induction coils have already been used for the same purpose, but they require the circuit to be more perfectly insulated, and it is impossible to properly test the circuit with a galvanometer.

115. **Application to domestic uses.—Saturn's "tinder box."**—Figures 29, 30 & 31 represent a novel arrangement, in which one of the secondary cells is placed, easily permitting a light to be obtained in laboratories and for domestic purposes.[1]

This apparatus, which we have designated, according to the traditions of the ancient chemists, by the name of

Fig. 29.

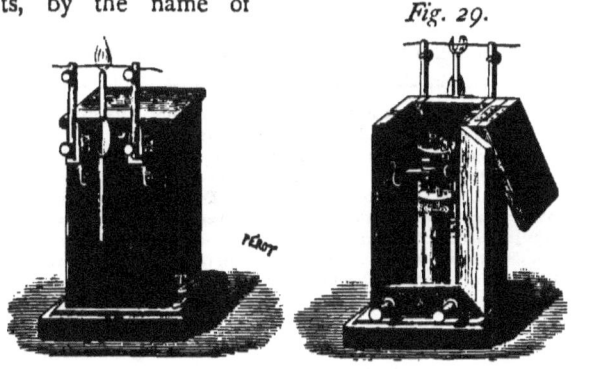

Fig. 30.   *Fig. 31.*

---

(1) Comptes rendus, t. LXXVII, p. 466. 1873.—Les Mondes, t. XXXI, p. 747. 1873.

Saturn's "tinder box," is composed of a small secondary cell of well formed lead plates enclosed in a box, on the bottom and sides of which is arranged a system of connections, so as to enable a platinum wire to be raised to incandescence, by the simple pressure of the finger upon a metal tongue, in order to light any inflammable substance, such as the wick of a candle, a mineral oil lamp, or gas, &c.

The apparatus is charged, and preserves its charge constant by being placed against two metal plates fixed upon the sides of a box containing a battery of three Daniell or Callaud cells (32).[1]

This battery may also be placed at a distance, if more convenient, and its poles connected by wires to a little board upon which the terminal springs serving as connectors are fixed (33), against which are placed the metal plates of the "tinder box," for charging.

Fig. 32.

---

(1) Léclanché cells which are so convenient to use in many cases cannot be advantageously employed in this instance. They would run down too quickly, as the circuit is nearly always closed in order to keep the apparatus charged. Besides, their E.M.F. diminishes considerably by prolonged closing of the circuit, so the secondary cells become less highly charged with three of these elements than with three Daniell or Callaud elements. Callaud cells are to be preferred, because, the sulphate of copper, remaining in a dense layer at the bottom of the jars, is less rapidly reduced by the zinc than in the Daniell elements of porous jars in which the sulphate of copper, filtering through this jar, becomes directly exposed to the reducing action of the zinc without any advantage to the current.

Three weeks may pass without renewing the sulphate of copper in a Callaud battery used for maintaining the "tinder box," whereas in the Daniell battery it must be renewed every week.

The Callaud battery only requires more time allowed it to begin working, and to attain its full E.M.F. the first time it is set up.

There is the great advantage in this arrangement of being able to charge the apparatus, and to remove it charged, without attaching or detaching any connecting wire. Any mistake in the direction of the primary current is also prevented. The positive pole of the secondary cell corresponding with the terminal C (fig. 31) must always come into contact with the same terminal of the primary battery, whichever way the apparatus happens to be turned.

When the secondary cell has been charged by a prolonged action of this battery, it is only necessary to press the finger upon the metal tongue arranged to close the secondary circuit in order to make it work. The platinum wire is thus raised to a sufficiently high temperature to instantly set fire to any combustible substance[1] (fig. 29).

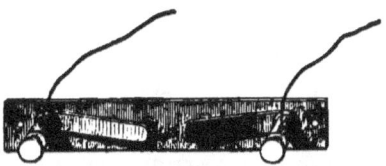

Fig. 33.

With the store of electricity preserved in this little secondary cell, after being charged to its full extent by a weak but prolonged current from the battery, a hundred lights may be consecutively obtained. It follows that it is not necessary to keep the secondary cell constantly fully charged by the action of the battery, and the object of being always in connection is to economise the current from the battery when it is thought that the secondary cell, not having been exhausted by a considerable number of successive discharges, might still produce many lights without being recharged.

---

(1) It is essential that the wick of the taper or small candle be traversed by the platinum wire, because if the wick happened to be too far below it would not light so easily, and besides, if it should catch fire, the platinum wire would be immersed in the hottest part of the flame, and, being at the same time raised to a white heat by the passage of the current, it might melt.

The lighting of a candle, by means of platinum wire raised to a white heat, is produced noiselessly and more instantaneously than by any other means. As the incandescence of platinum wire does not in any way alter the composition of the air, there is no smoke or smell of suffocating gas, like that which takes place with sulphur or chlorates. One need not fear risk of fire, or poisoning by phosphorous. We may finally take this means of lighting as very economical; for, on the one hand the secondary cell requires no expense at all for maintenance, the lead and the solution being put in once for all without ever requiring renewal, and, on the other hand, it suffices, in order to keep the weak current of the charging battery in action, to add from time to time a few crystals of sulphate of copper, of which the consumption is extremely small compared with the great number of times the light may be obtained.

The "tinder box," once charged, may be carried away, and, in consequence of these cells having the property of preserving their charge, a considerable number of lights may be obtained without putting it back in connection with the primary battery.

Fig. 34.

116. Figure 34 represents another novel arrangement of the same apparatus, making it a kind of electric candle-stick. The terminals, between which the platinum wire is held, and the little candle are fixed on a separate board in connection with small vertical plates of metal.

By placing these small plates against corresponding plates upon the secondary cell, the incandescence of the platinum wire, and consequent lighting of the candle, is instantaneously produced, and the latter may be as easily carried away as any ordinary candle-stick.

117. "Saturn's tinder box" may also be combined with electric bells, so as to work from one primary battery without in any way impeding the action of the bells, by placing it in direct communication with the two poles of the battery, and thus forming a shunt circuit.

It would appear, during the charge of the secondary cell under

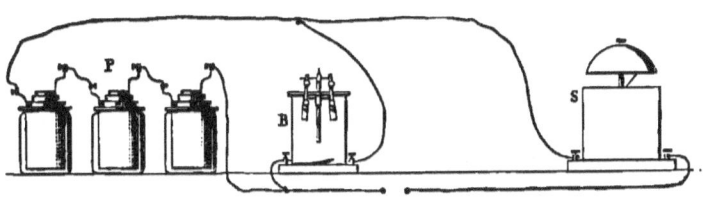

Fig. 35.

the action of a battery in the circuit of which are placed one or more bells, that this apparatus must absorb all the current and prevent the bells working, but as the lead plate secondary cell acquires under the influence of the current a high intensity, the result is that it does not act like an inactive derived current but even contributes towards working the bells. Especially if the battery itself happens to be too weak to make the bells ring, the secondary cell becomes capable, by means of the power stored, of putting them in action. It acts in this case like an accumulator of work done, a sort of electric fly-wheel.[1]

---

(1) Comptes rendus, t. LXXVII, p. 466. 1873.

Besides, as bells only work in an intermittent manner, allowing from time to time, sufficient intervals for the secondary cell to be charged, the cell does not quickly become exhausted, even if the bells work continuously; for, in consequence of this store of electricity, a secondary cell well charged will work one or more electric bells continuously for more than an hour.

The two kinds of apparatus may even work simultaneously without injuring one another. Thus, at the same time, the candle may be lighted and the bells sounded. This happens from the fact that the secondary cell is in the shunt circuit, and as the platinum wire possesses a considerable resistance, part of the current passes through the bell circuit.

118. The same apparatus may be applied to lighting gas, and the more easily because gas does not require the incandescence of so large a platinum wire as a wax or stearine candle. Consequently it may be effected at a considerable distance, and if it is a question of lighting a large number of jets simultaneously, recourse may be had to batteries composed of a great number of secondary cells, just as used for firing mines.

119. **Application to Electric Breaks for use on Railways.**—We have recommended the use of the secondary cells above described, wherever a current of great intensity is required for a very short time, to produce either calorific or magnetic effects by means of a feeble source of electricity.[1]

M. Achard used them recently with success for working his electric breaks, which require at a given moment the passage of a strong current through a series of electro-magnets wound with thick wire.

---

[1] *Recherches sur les courants secondaires et leurs applications.* Annales de Chimie et de Physique, 4e série, t. XV, p. 20, 1868.

The secondary cells are kept charged by a primary battery of three Daniell elements, the current of which they accumulate, as in the preceding applications (89, 109, 217).

The accumulated force is then spent, in an instant, in the form of magnetic work.

The primary battery remains in constant connection with the secondary cells. Thus, as we have above explained (46), it combines its feeble action with that of the secondary cells during the discharge, and again acts upon the secondary cells by charging them as soon as the circuit of discharge is interrupted.

120. Application to the eudiometric analysis of the atmosphere of mines.—In all cases where a calorific effect of short duration is wanted these cells may be advantageously employed. It is thus that M. Coquillon has made use of them for heating a palladium wire, and determining the combination of air and proto-carbonised hydrogen in his fire damp detector.[1]

121. Application to the production of Luminous Signals.—The secondary batteries described (96), are able to produce a voltaic arc of several seconds duration and of very great intensity by employing a sufficient number of secondary cells, charged for a few minutes from two Bunsen elements, and we have noted the use that might be made of them in certain instances for the production of luminous signals.[2]

Although this idea has not yet been put into practice, we feel sure that it would be of great service at sea, or along the coasts, because the inconvenience and cost resulting from the use of batteries become considerably reduced when it is only a

---

[1] Comptes rendus, t. LXXXV, p. 1106, 1877.
[2] Brevet du 29 Février, 1868.

question of setting up two Bunsen elements in order to obtain at any moment an electric light equal to that which eighty or a hundred of these elements will give direct.

Mons. A. Niaudet[1] has taken a special interest in this application, and recommended the use of the Gramme machine[2] for charging secondary batteries intended for the production of luminous signals, by means of mechanical force, which signals would often prevent collisions at sea.[3]

Mons. J. Morin[4] has made several experiments with the same object by means of a secondary battery of fifty large cells worked by a small magneto electric-machine, having only eight coils. This battery could fuse, in discharging, an iron wire 2·20 metres in length and 0·001 metres in diameter. Mons. Morin has been occupied in the construction of a special electric lamp for producing the voltaic arc under these conditions.

122. **Application to the production of the Electric Light in special cases.**—In certain cases, where a bright light is required for several minutes only—in projection experiments for example, or for any other study—the battery of six large cells above described (100) answers the purpose.

Mons. Reynier's electric lamp allows of obtaining, some good results, even with this low tension.

As this lamp possesses considerable resistance if carbons of small diameter are used, the expenditure of the electricity stored

---

(1) V. La Nature, 27 Juin, 1874.
(2) The Gramme Machine has been used for several years in the workshops of Mons. Breguet for forming and putting secondary batteries into working order.
(3) It is known that Mons. Trève has studied in a special manner, and by other means, the solution of this important question.
(4) Comptes rendus, t. LXXXI, p. 435, 1875.

during two hours in the secondary battery is comparatively slow, and, consequently, the light may last for a quarter of an hour.

By making use of a larger number of these big cells, connected together in two or three batteries, a very bright light could be obtained, long enough to be of use in certain cases.

123. Application to the sub-division of the Electric Light.—The lead plate secondary battery composed of forty small elements, designed to produce continuous effects, of which we have previously spoken (101), could, by giving a sufficient rapidity to the commutator, supply a continuous voltaic arc, of proportionately reduced power it is true, but obtained by means of only two ordinary Bunsen cells. Two large size cells would keep several similar arcs going by working through a considerable number of similar small batteries, and there would be thus obtained a solution to the problem of sub-division of the electric light, by means of secondary currents.

This solution, which we pointed out some twelve years ago,[1] appears rather complicated. Still, it is not impossible to carry out; for commutators constructed like those described further on (part V.), do not require a high E.M.F. in order to keep them going; they could be arranged so as to turn altogether like the bobbins of spinning looms, and besides, secondary batteries employed for this transformation could be put into quite a small space, as they need have but a very reduced surface.

Mons. W. Lermantoff has devoted himself to experimental work upon this subject.[2]

---

(1) Brevet du, 27 Avril, 1868.
(2) Journal de Physique, t. 258, 1876.

124. **Physiological Effects produced by Secondary Batteries.**—The E.M.F. of each secondary element of lead plates being, as already noticed (80), fairly strong, secondary batteries of twenty or forty cells suffice to give very powerful physiological effects. These effects could be utilised in therapeutics, by using secondary cells of very small surface, so as to avoid any effects of temperature. The secondary battery for continuous currents, of which we have already spoken (101), would be suitable for this purpose; even batteries intended for only temporary work could be employed, as the discharge of these would last long enough to produce an effective action, in consequence of the high resistance of the human body.

125. **Various Applications.**—Secondary cells and batteries as above described, may be, generally, applied in every instance where, at a given moment, a powerful electric effect, either of quantity or tension, is required.

Such is the case, for example, when it is a question of transmitting the time of day simultaneously to several different places by means of a current through a number of wires.

These apparatus may be of great use in scientific research, as we shall see further on (Part III).

Mons. Thore, of Pau, in 1875, employed a light furnished by a small battery of twenty cells for spectroscopic experiments.

Mons. Guérin in 1875, used these cells in electro-chemical gilding and silver plating, in certain cases where a very strong current, of quantity, was required for a short time.

126. Other applications are recommended, based upon the results of our research in voltameters.

Firstly, in 1860, the substitution of lead electrodes for those of platinum, used by Jacobi to produce counter polarisation currents, for preventing delays in the signals upon certain telegraphic lines which were imperfectly insulated.

Secondly, in 1865, the use of lead anodes instead of platinum for electro-typing.

127. A fact to which we have drawn attention when specially studying copper wire voltameters ;[1] *viz.* the formation of a very fine point at the extremity of the positive electrode, has been the subject of some experiments by the engineer, Cauderay, of Lausanne, for the pointing of pins by electro-chemical means.

Apparatus made upon this principal was exhibited at the International Exhibition of 1867, and, if this application has not been followed up since the death of Cauderay, it no less deserves to be noted and again taken up, on account of the unhealthiness resulting from the mechanical process of pointing pins, &c.

128. The phenomenon we noticed in 1859,[2] and which we have previously described (23), *viz.* the almost complete cessation of a current traversing a voltameter of alluminium electrodes, by reason of the insolubility of the alluminium formed upon the positive pole, has recently furnished Mons. Ducretet with an idea for some ingenious appliances in telegraphy.[3]

---

(1) Bibl. univ. de Genève, t. VII, p. 332, Avril, 1860.

(2) V. Comptes rendus, t. LIX, p. 610, 1859.

(3) V. Bulletin des séances de la Société francaise de Physique, p. 17 et 40, Janvier-avril 1875.

THIRD PART.

Effects produced by Electric Currents of high Tension.

CHAPTER. I.

Luminous Sheath.—Luminous Liquid Globules.—Globular Flames.—Voltaic Brush.—Luminous Figures.—Moving Spark.—Bunch of Aqueous Globules.—Jets of Steam.—Electrified Liquid Vein with Spiral Motion.—Electric Bar.—Voltaic Pump.—Electro-Silicious Light.—Crowns, Arcs, Rays, and Undulating Movements.—Electro Dynamic Spirals.—Crater-like Perforations.

129. The secondary cells above described (96—98) have permitted of the study of phenomena produced by electric currents of high tension, and particularly those which appear in the passage of these currents through liquids.[1]

---

[1] Research in Phenomena produced by Electric currents of high tension in liquids. Comptes rendus, t. LXXX, p. 1133, 5 Mai, 1875.

Some phenomena of this kind have already been made the subject of study with ordinary batteries by Davy, Hare, Makrell, Grove, Gassiot, de la Rive, Wartmann, Despretz, Fizeau & Foucault, Quet, Maas, Van der Willigen, &c ; but the necessity of having to set up a powerful battery for their observation was such a difficulty that they have never been deeply analysed. The currents supplied by secondary batteries are, it is true, but temporary; they have, nevertheless, sufficient duration to enable us to follow in all their details the effects produced by the passage of electricity through imperfectly conducting substances, such as the solution in voltameters; besides, the experiments may be renewed by recharging the apparatus, and the intensity of the current slowly diminishing as the discharge continues, far from proving inconvenient, successively shews to the observer a series of different phases, which would escape remark with a constant current, or would require continual modifications in the elements of the battery.

The study of these phenomena also presents so much the more interest, because, at this point are found the two powers which exercise the most direct influence over the elements, viz: electrical force and chemical force, where a solution for every problem in human industry appears to be found.[1]

In fact, by following the passage of currents of varying intensity through liquids, we watch, so to speak, the contest between "the electric flow" and molecular attraction, added to chemical affinity, tending to hold united the metal molecules of the electrodes, or the elements of the liquid body contained in the voltameter. If "the electric flow" possesses great intensity, mechanical and

---

(1) Dumas. Bulletin de la Société d'Encouragement, t. XII, p. 153, 1866.

calorific effects predominate ; the molecular attraction is first overcome, and the electrodes are disintegrated, fused, or vapourised. If the intensity is somewhat less, the electrodes become the seat of luminous phenomena, produced by the vacuum and rarified vapours around; the solution, hardly moistening the electrodes, is scarcely decomposed. If the intensity of the current still decreases, most of the calorific and luminous phenomena disappear, and chemical decomposition is exhibited; and as, on the other hand, the current then passes more thoroughly in the liquid, its intensity seems greater throughout the circuit. This may be shown in a very striking manner by the following experiment.

130. **Experiment upon the "luminous sheath" with the current decreasing in intensity.**—The current from the secondary batteries, each composed of twenty pairs of lead plates, is discharged through a voltameter V, of water acidulated by sulphuric acid, and platinum wire (36). The positive wire only is

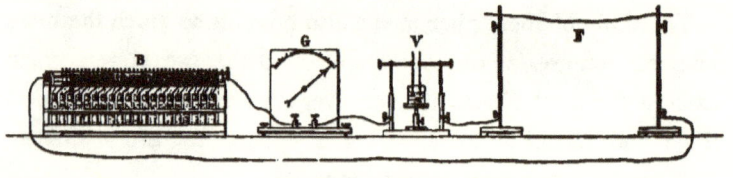

Fig. 36.

first dipped in the liquid. A galvanometer G, is already arranged in the circuit and a platinum wire F, stretched through the open air, about 80 centimetres long and 0·1 m.m. in diameter, is also placed in the circuit. Immediately the negative platinum wire is dipped in the solution there is seen upon this wire, without any noticeable liberation of gas, a *luminous sheath* such as has been observed with ordinary batteries by the above-mentioned scientists. The positive wire, on the other hand, sets free but a very small quantity of gas. The galvanometer only shews a slight deviation, and the platinum wire stretched through the air is not reddened. But if we leave the experiment alone, in two or three minutes, as the E.M.F. of the secondary battery falls, the luminous sheath disappears, and an abundant liberation of gas takes place suddenly at both poles; the galvanometer shews a strong deviation, and the platinum wire is at the same moment rendered incandescent throughout its length.

131. The various phenomena which may be produced with different metals and solutions, according to which pole is first dipped, and which have been studied with great care by Van der Willigen,[1] by means of a Bunsen battery of forty elements, are easily re-produced with a secondary battery of forty cells, and we believe that the following rule holds good for these phenomena under the conditions in question, viz: the electrode first dipped in the solution, or the one which presents the larger surface submerged, gives its sign to the liquid in the voltameter.

132. **Change of colour in the "luminous sheath" according to the intensity of the current.**—In proportion as the battery is discharged and the intensity of the current decreases, we have noticed that the colour of the luminous sheath formed round the negative electrode gradually alters; it passes successively from white to blue and violet, and towards the last, a few seconds before the liberation of gas appears, it is reduced to a few brilliant points of a reddish purple, which surround the end of the electrode.

At first we thought there might be some possible connection between the intensity of the electricity in play and the refrangibility

---

(1) Annales de Poggendorff, t. XCIII, p. 285.

of the light produced; but later experiments with currents of greater intensity having more clearly shewn the nature of the phenomena which take place around the electrodes, these changes of colour may be explained in the following manner:—The luminous sheath is nothing else than an envelope of rarified and incandescent gases formed round the electrode, and of equally rarified and incandescent vapour from the solution of the voltameter.

What is the nature of these gases? In consequence of the very high temperature produced round the electrode with a current of great intensity, the water is partially decomposed around the same pole, as proved by Grove, and as our own researches have verified. There is then hydrogen, oxygen, and sulphuric acid, or sulphurous vapour, round the electrode, when the liquid is water acidulated by this acid. It can also be easily understood that the nitre arising from the atmosphere may be held in solution by the liquid. All these elements are rarified and luminous, and the colour of the light necessarily participates in the mixture. A white tint predominates, probably arising from the relative abundance of sulphurous vapour given out by the liquid. If salt water is used the sheath emits a brilliant yellow light, due to the excess of sodium.

But, as the intensity of the current decreases and the heat diminishes, these dissociations become less complete, the proportions of the various products become modified, and consequently the colour varies.[1] When the current is reduced to a very feeble

---

[1] We also notice in Geissler tubes submitted to a prolonged action of an induction or high tension current, similar changes in the colour of the light emitted; but in that case the alterations are not caused by variations in the intensity of the current; they are permanent and are due to modification in the gaseous matters contained in the tubes, the limited quantity of which is not renewed.

intensity, the calorific action lessens and the point is approached where electrolysis of the water is produced in the ordinary manner; hydrogen begins to dominate alone at the negative pole, and, if the heat caused by the current is still sufficient, the latter period of incandescence is maintained for a few moments.

Hence this purple colour in the light, which finally appears at the extremity of the negative electrode; for we know that such is the colour proper to incandescent hydrogen enclosed in a narrow space. Now, the luminous sheath here becomes so much the more compressed by the vicinity of the liquid, which the calorific effect of the current has further diminished.

133. Batteries of from two hundred to eight hundred secondary cells, used in studying electrical effects of high tension.—In order to observe effects produced by electric currents of very high tension, we successively connected up batteries of two hundred to eight hundred secondary cells, the discharge of which, for the first few moments after the action of the primary current, equals that of from three hundred to twelve hundred Grove or Bunsen elements.

The E.M.F. of each secondary cell, immediately upon breaking the primary current, is, we know, equal to one and a half Grove or Bunsen elements, as above shewn (80). This E.M.F., it is true, undergoes some fall when the secondary circuit is not immediately closed upon the primary circuit being broken; but, in spite of this fall, it still remains higher than either a Grove or Bunsen cell (81).

The resistance of the cells composing these batteries is, on the other hand decidedly lower than that of the Bunsen cell of ordinary dimensions, in consequence of the very close proximity of the lead plates, and in spite of their limited surface (two square decimetres). This resistance is hardly equal to three metres of copper wire one m.m. thick (83).

The result is that each of these small secondary cells is capable of producing, when well charged, a calorific effect sufficient to raise a platinum wire, ·3 to ·4 m.m. in diameter by five cents. in length, to a red heat. With two secondary cells only we have been able to redden a platinum wire of this diameter and ten metres long.

This incandescence is certainly of very short duration in consequence of the small surface of each plate; but if the discharge take place through circuits of higher resistance, for instance, if we study its effects upon the surface of a liquid, the expenditure of current is much less rapid, and, with a single discharge, we have often been able to repeat more than twenty experiments without completely exhausting the charge in the battery.

134. Figure 37 represents an arrangement of four hundred secondary cells divided into ten batteries, each of forty pairs. These batteries are of the same shape as those already described in figure 20 (98); but they are composed of double the number of cells.

In our later experiments, made with eight hundred secondary cells, a second series of batteries exactly similar was arranged in another room, and the current connected by wires to that of the first series.[1]

These batteries, coupled up first in parallel by means of commutators, only require two or four Bunsen cells in order to be

---

[1] Still more recently we arranged these batteries in steps upon shelves, so as to allow of their being placed in a still smaller space.

Fig. 37.

charged all at once, and these Bunsen cells may be placed outside a window, upon the sill, so as to avoid the acid fumes. When the batteries have not remained too long without working, a few hours are sufficient to charge them.

Then, by turning the commutators, all the secondary cells may be connected in series, and the charge resulting from the chemical work accumulated during several hours by the two or four Bunsen cells may be expended at will, either in a few seconds, or in any longer time.

Experiments are most often made in the dark, so that the details of the luminous phenomena produced may be studied. The voltameter is shewn at the moment when the electric current happens to act upon its surface. The water vapour may still be seen to be set free above the liquid, after the powerful calorific effect produced by the passage of the current.

Some rheoscopes of platinum wire, the same as that shewn in figure 24 (102), are placed upon the tables, and serve to test the condition of the secondary cells, in which some accident may occur, as above explained.

Other large rheoscopes, with a long platinum wire stretched between the terminals, allow of the separate examination, if necessary, of the condition of each battery.[1]

---

[1] When it is required to put all the batteries into work, the experiments are rather delicate and need care in preparation, on account of the multiplicity of secondary cells and great number of metallic connections necessary.

They are also not free from danger in carrying out, for the discharge of such currents combining, at one and the same time great quantity and potential, can produce very violent shocks upon the human frame. For three years we were fortunate enough to avoid anything of this kind; but upon the occasion of an experiment made lately, in order to charge the rheostatic machine, which will be described further on (part V.), having involuntarily touched the naked ends of the wires from a series of six hundred secondary cells, we instantly felt, not only a very strong shock, but the sensation of burning throughout the body, rising as far as the neck, causing us to cry out, and considerably frightening the people standing round. All the same, this accident had no unpleasant consequences. But it might not perhaps have been so if the eight hundred secondary cells had then been in action. Shocks given by induction coils never seem to produce the same effect.

135. Luminous Liquid Globules.—If a battery of two hundred cells be put in connection with a voltameter of water acidulated by sulphuric acid, or of salt water, so that the positive wire alone is immersed to begin with, the approach of the negative wire towards the liquid causes the fusion of this wire, or its vaporization, with a kind of explosion, and a flame variously coloured, according to the kind of metal forming the electrode.

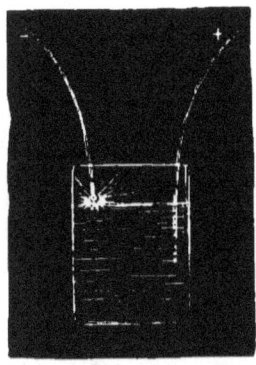

Fig. 38.

By diminishing the portion of acid contained in the solution of the voltameter, so as to avoid complete fusion of the metal, a continuous series of sparks, accompanied by a cracking noise, is produced, and these sparks continue for several minutes, decreasing gradually in intensity (Fig. 38).

But if, when the negative wire is immersed first in the solution (which ought to be by preference salt water

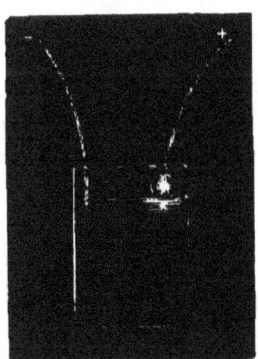

Fig. 39.

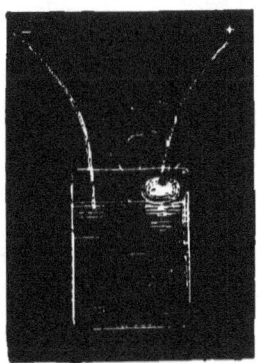

Fig. 40.

so as to avoid acid vapours, and slightly raise the resistance of the circuit), we bring the positive wire close to the surface the result is quite different.[1]

The wire is not melted and a small luminous liquid globule, accompanied by a curious noise, is seen formed at its extremity (Fig. 39). On gradually withdrawing the wire the globule increases in size, as if the liquid was sucked up by the electrode; it acquires a diameter of about one centimetre and takes at the same time a rapid spiral motion. In consequence of this motion it becomes flattened (Fig. 40),—elongated sometimes towards the negative wire, if it is near enough,—and is finally dispersed at the same time that a detonating spark is produced at the negative pole when this pole only dips a very little way in the liquid. The globule is again spontaneously formed at the end of the positive wire, and the same phenomena thus take place several times in succession in an intermittant manner.

136. The spiral movement does not always take place in the same direction, like the magneto-electric spiral motions described further on (158). It takes place sometimes in one direction and sometimes in the other. It often happens to go in the same direction a number of times running; but this may change without any apparent cause. It is a re-action spiral motion, similar to that of electric whirls, and is due to the flow of the electric current through the liquid.

The globule becoming nearly detached by its spherical form from the rest of the liquid, or only having a very small surface in contact with this liquid, the motion takes place in either direction,

---

[1] The voltameter is placed upon a support furnished with crooks, to which are fastened platinum wires in connection with the poles of the battery, so that they may be introduced with proper care into the liquid.

according to the position of the point upon the surface of the globule through which the main flow of current passes or the liberation of vapour is produced

The luminous appearance of the whole of the globule seems to arise from the bright light emitted at the point of contact with the rest of the solution.

The noise is due to condensation of the steam, which tends to form round the electrode, in the liquid. The intermittant spark, which appears at the negative pole at the instant the globule attains its maximum development, is explained by reason that the negative wire, at first dipped a little way into the liquid, soon becomes separated from its surface by the vaporization of the portion of liquid which forms the globule. The current is then broken for a moment, the liquid from the globule, falling back into the voltameter, re-establishes the contact, and the phenomena may be thus re-produced several consecutive times, spontaneously, until the secondary cells are exhausted.

137. As to the concentration of the liquid in this globular form, it would seem to be explained by the phenomenon of "suction," resulting from the flow of the electric current itself at the positive pole. We shall see further on a still more striking example of this "suction," when a current of higher tension is used, and when the space for liquid round the electrode is limited by being enclosed in a narrow tube (148) ("voltaic pump"). But, in this case, the liquid, having unlimited space, naturally concentrates into the form possessing the smallest possible surface, consequently becoming spheroidal.[1]

---

(1) This spheroidal form taken by a liquid, under the action of the calorific effect produced by an electric current, may be compared with that likewise shewn by liquids under the action of heat alone, when placed upon red hot surfaces, which have been studied by Boutigny. It is also the form taken by liquids when simply withdrawn by the action of gravity, as shewn by the experiments of Plateau.

138. Globular discharge.—Brush discharge and luminous figures produced by the discharge of a battery of 800 Secondary Couples.—For studying effects produced with a voltameter of distilled water, we have quadrupled the potential of the current, by uniting 20 batteries, each composed of 40 couples, forming a grand total of 800 secondary cells.[1]

When the current from the whole of this battery is passed through distilled water, there is exhibited at first, in a more intense degree, a phenomenon somewhat similar to that observed by Grove, with 500 elements of his nitric acid battery. The positive electrode being previously immersed in the distilled water, by bringing the negative platinum wire close to the surface of the water, and raising it immediately, there is obtained a yellow flame, nearly spherical, about 2 centimetres in diameter (fig. 41). The platinum wire, from one to two millimetres in diameter, melts rapidly and keeps for several moments in a state of fusion, at a height of from 14 to 15 millimetres above the liquid.

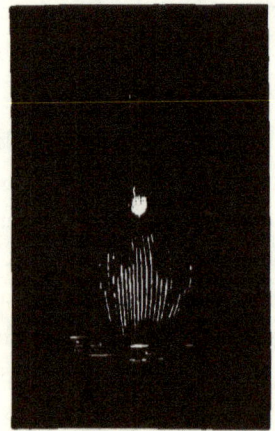

Fig. 41.

This flame is caused by rarefied incandescent air, by the vapourised metal of the electrode, and by the elements of the decomposed water; spectral analysis clearly shows the presence of hydrogen.

If, in order to prevent fusion of the medal, the intensity of the current be diminished by interposing a column of water in the circuit, the spark appears in the very clearly defined form of a

---

(1) Comptes rendus, t. LXXXV, p. 619, October, 1877.

little globe of fire, from 8 to 10 millimetres in diameter (fig. 42).

By slightly raising the electrode, this globe takes an ovoid form; blue luminous points, their number continually varying and arranged in concentric circles, appear on the surface of the water (fig. 43). Rays of the same colour soon shoot forth from the centre and join these points (fig. 44).

From time to time, the rays take a gyratory motion, sometimes in one direction, sometimes in another, describing spiral curves (fig. 45 and 46). Sometimes the points and rays all disappear from one side, and varied curves, formed by the movement of those which remain, become visible on the surface of the liquid. Finally, when the rapidity of the gyratory movement increases, all the rays vanish, and nothing more is seen but the blue concentric rings (fig. 47). The rings are the last phase of these transformations, which are very curious when followed with the

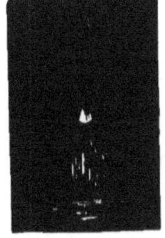

Fig. 42.

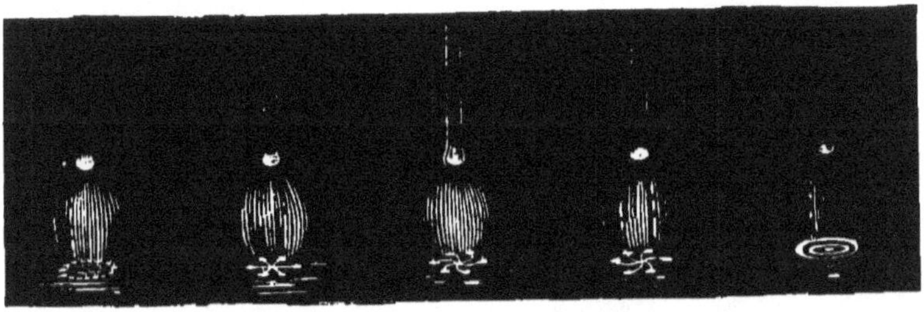

Fig. 43.   *Fig. 44.*   *Fig. 45.*   *Fig. 46.*   *Fig. 47.*

naked eye or with a glass, and constitute quite an electric kaleidoscope.[1]

---

[1] These phenomena may be compared with those observed by M. Fernet with induction currents; Comptes Rendus, 1864; they also greatly resemble those resulting from the fall of liquid drops on a level surface, studied by MM. Helmholtz, Thomson, Maxwell, Tait, Rogers, Worthington and Trowbridge.

139. The production of these figures is explained by the excessive mobility of the arcs or luminous fibres which compose the ovoid light, between the water and the electrode. By carefully examining this particular form of spark, it is discovered, that in reality it is a kind of tuft or voltaic spray, similar to that in static electricity, but more complete, by reason of the greater quantity of electricity in play. These luminous fibres, being in a continual state of agitation, the points where they meet the surface of the liquid constantly change, and form the rays which are seen. Their gyratory motion arises from the reaction due to the electric flow. As to the rings, they visibly form under the eye of the observer, by the motion of the blue points becoming more and more rapid, and by the continuance of the impression upon the retina.

140. When the metal electrode is positive and the distilled water negative, the spark again takes outwardly an ovoid form, but the middle is traversed by a cone of voilet coloured light.

When two metal electrodes are used, a luminous spheroid is obtained, the interior of which is traversed by a brilliant streak of light. This appearance corresponds with the streak and the halo of the spark from induction currents; only here, the halo occupies more space, in consequence of the greater quantity of electricity. In short, if the length of the column of water interposed be much increased only an arc or rectilineal ray of light is obtained.

It is not necessary, in these experiments, to bring the electrode into contact with the water, to cause the passage of the electric current. The tension of the batteries, although the couples which compose them are not insulated in any special way, is great enough to make the spark strike spontaneously at the distance of about one millimetre above the liquid.

141. If, instead of leaving the electrode fixed on the surface of the voltameter, during the flow of the electric current in the form of these sparks or globular brushes, one of the wires serving as electrode be suspended at a considerable height, and enough weight and length given to it to cause it to oscillate like a pendulum on the surface of the liquid, or over a conducting plate, without perceptibly changing its distance from this surface, the little globule of fire, produced at the extremity of the wire, naturally follows the movement of the electrode, and, when working in the dark, nothing but the globule of fire is seen to move on the surface of the liquid. Further on, we shall refer to this experiment (part IV.) in order to explain certain appearances in natural electric phenomena.

142. **Wandering electric sparks.**—The electric spark in this globular form, resulting from the action of a great quantity of electricity upon ponderable matter, may be excited to a progressive motion, by itself, without it being necessary to move either electrode.

This is the result of a more recent experiment we have made[1] by using the apparatus described in the Fifth Part under the name of Rheostatic Machine.

Although this experiment does not necessitate the use of a voltameter, we will here give a description of it, because it relates to the globular forms of electrified matter of which we have just quoted several examples, and we will compare this experiment with the preceding ones, to explain, by analogy, the slow progression, in certain cases, of globular lightning.

If the two poles of the secondary battery of eight hundred couples are put in connection with the armatures of a condenser,

---

[1] Comptes rendus, vol. LXXXVII, p. 325, August 19th, 1878.

the insulating plate of which is formed by a sheet of mica, this condenser becomes charged like a Leyden jar, and can give, when discharged, a spark of the nature of static electricity.

But if, by chance, the mica plate has some very thin place or crack, made when it was cut out, it is spontaneously pierced at this point by the action of the current from the eight hundred secondary cells, the same as the glass of a Leyden jar over charged by an electric machine.

A remarkable phenomenon is then presented to view. In consequence of the great calorific power of the electricity in play in this experiment, the spark which has struck across the condenser, between the two armatures, is not instantaneous like that of static electricity, but, as it is accompanied by fusion of the metal and even of the insulating material of the condenser, it forms a small and very brilliant luminous globule which moves with a peculiar noise, and slowly traces, on the tinfoil of the condenser, a deep track, sinuous and irregular.

Figure 48 represents a true copy of that portion of the surface of a condenser where the phenomenon occurs. The spark appears at A, branches out soon at B, as far as C; there it disappears, to reappear immediately at the point B, with such rapidity and in an interval of time so hardly appreciable, that it seems to have made a bound; next, it goes towards D; there it forms a new branch which stops at E, reappears at D, continues its way towards F, and so on. Sometimes, as in the present case, the spark appears again farther on, at the point Q, detached from the principal track, to stop next at R, and the phenomenon only ceases when the mica plate presents no other part thin enough to be pierced. In other cases the spark remains some time stationary round the same point, other times again, one of the branches is immoderately prolonged, and describes, on the entire surface, outlines similar to those on a geographical map. A tube of distilled water is

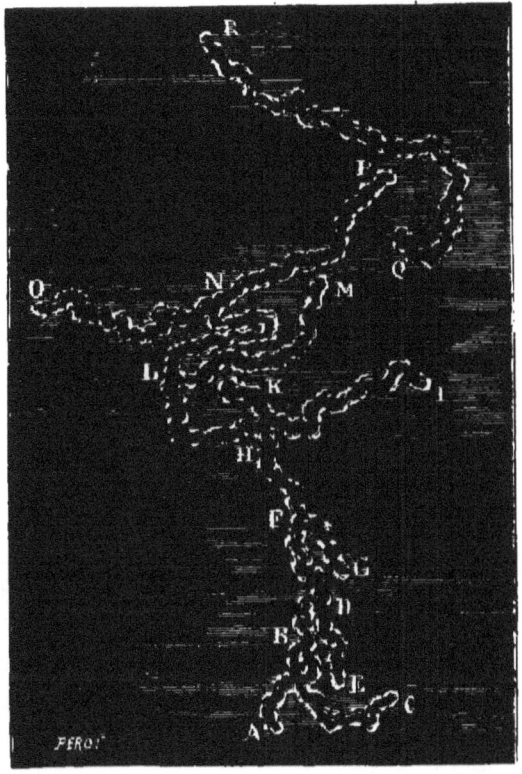

Fig. 48.

previously interposed in the circuit of the secondary battery, to prevent too intense calorific effects and deflagration of the whole condenser.

Whilst the phenomenon is happening, the points through which the spark will pass cannot be foreseen; nothing is more curious than the motion of this little shining globule, which is seen slowly making its way, choosing the points towards which it must go, according to the greater or less resistance of these points in the insulating plate.

The condenser is perforated by the passage of the spark, and the tinfoil forms a double row of melted beads around the edges of the consumed mica.

Fig. 49.

143. **Sheath of aqueous globules.**—Taking again the solution of salt voltameter in which the current from a battery of two hundred cells produces at the positive pole a luminous liquid globule; if the tension of the current happens to be doubled, by employing a battery of four hundred cells, the effects are completely changed.

There is then obtained, by the immersion of the positive wire, instead of a single globule, a sheaf of innumerable ovoid globules which succeed each other with excessive rapidity, and are projected to the distance of more than a metre from the jar in which the experiment is made (fig. 49). It is a kind of *pulverization* of the water into small drops, produced by the electric discharge.

The spark in this case appears at the surface of the liquid, in the shape of a crown or many pointed halo, from which spring out aqueous globules.[1] The metallic form of electrode is not necessary to obtain this effect; a fragment of filter paper,

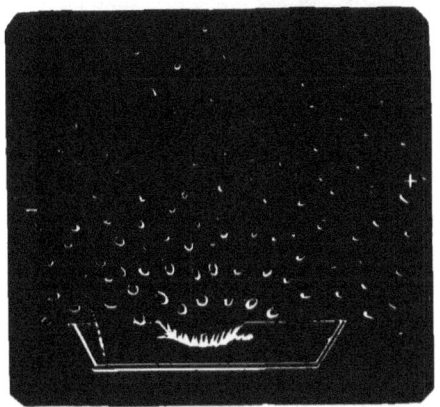

Fig. 50.

moistened with a solution of salt connected with the positive pole, also produces this phenomenon (fig. 50).

144. Jets of Vapour.—If the current, instead of encountering a deep layer of liquid, finds only a moistened surface, such as the sides or inclined bottom of the jar, the calorific effects predominate, the halo is more brilliant and the water is quickly vaporized.

The action of the current differs then according to the resistance opposed to it, and here is found another example of reciprocal substitution of the heat and mechanical work resulting from the electric shock. When the work, represented by the

---

(1) Comptes rendus, vol. LXXXII, p. 314, January 31st, 1876.

violent projection of the liquid, appears, there is neither heat nor vapour developed, and when no visible work is accomplished,

Fig. 51.

when the liquid is not thrown about, heat is generated and vapour escapes.

145. The formation of these luminous tracks, accompanied by jets of vapour, is *intermittent*. Each time, in fact, that the electrode in contact with the moistened surface has vaporized the small drops of water which surround it, the current is for an instant interrupted, but another portion of the liquid mass which moistened this surface flows in immediately, and the phenomenon begins again, thus appearing intermittently, until the voltaic discharge be exhausted.

146. **Electrified liquid vein—gyratory motion.**—If a vein of saline water is made to flow from a funnel in connection with the positive pole of the same battery (400 secondary couples), into a basin in which the negative wire is already immersed, and beneath which is placed an electro-magnet (fig. 52), there may be perceived, after closing the voltaic circuit, a luminous

thread, accompanied by some brilliant points, in the lower part of the vein; crackling sparks spring forth at its extremity, vapour escapes from the water, and the liquid which surrounds the lower part of the vein makes a gyratory movement *in the opposite direction to that made by the hands of a watch, if the pole of the electro-magnet be north, and in the same direction, if that pole be south.* The movement is rendered visible by light substances spread on the surface of the liquid.[1]

Fig. 52.

If the vein be shortened so as to prevent any break in its lower part, the electric luminous signs disappear almost completely; the liquid nevertheless becomes heated, as shown by a slight vapour, and the gyratory movement becomes more clearly defined and more rapid. By again prolonging the vein, the luminous electric manifestations re-appear as before.

---

(1) Comptes rendus, vol. LXXXII, p. 220, January 17th, 1876.

147. Electric Bar.—By placing the positive electrode against the sides of a jar of saline water in connection with the negative pole, there may be observed, besides the luminous tracks and abundant jets of vapour, a violent eddy in the liquid, forming a

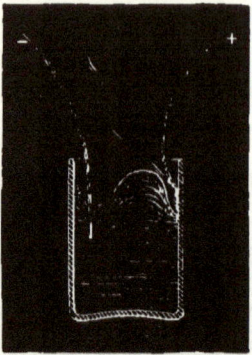

Fig. 53.

kind of electric *bar*, which raises the water to the height of $1\frac{1}{2}$ centimetres above its natural level (fig. 53). If the currents meet at certain points with inequalities of resistance, it may divide and give rise to two or three aqueous hillocks, as indicated in figure 54.

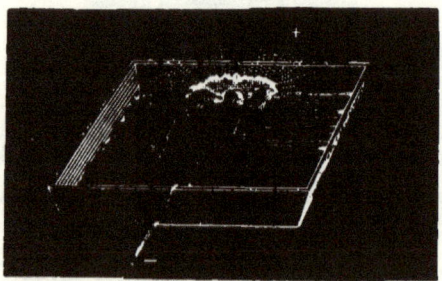

Fig. 54.

This phenomenon is another result of calorific effect, produced by the current upon the moistened surface which it encounters.

The liquid is repelled by the pressure of the vapour suddenly developed by the current at a fixed point.

This effect may be compared with the breath or draught produced by an abundant flow of static electricity. Only, in the latter case, if the tension be greater, the quantity of electricity is much less; also, if a current of static electricity produces an effect of this kind only upon air, it could not act in the same way upon a mass of liquid.

148. **Voltaic Pump.** — Remarkable effects of *suction* may also be produced by the electric current. If the positive wire be placed in a capillary tube, leaving however about half a centimetre free at its extremity, immediately the electrode tube is immersed in the salt water, the liquid is seen to rise with extreme rapidity, to a height of from 25 to 30 centimetres and fall again in a sheet of brilliant sparks and jets of vapour (fig. 55.)

A voltaic pump is thus formed, in which the vacuum results from the production and the condensation of the vapour around the electrode.

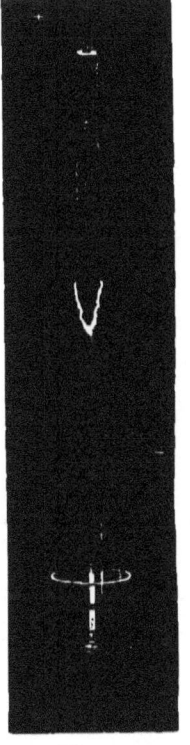

Fig. 55.

149. The luminous ring which accompanies the fall of the liquid from the upper to the lower part of the tube, and reappears again spontaneously and intermittently in the upper part, only to again descend, constitutes one of the most brilliant and curious effects that we have observed with electric currents of high tension.

This phenomenon is explained as follows: the liquid sucked up constitutes a prolongation of the negative electrode itself, formed by the liquid of the voltameter (131). Calorific and luminous phenomena should then be seen at the extremity of the liquid sheet, falling again from the top of the tube, especially if the glass, already moist, offers a certain conducting power, thus establishing an outer connection with the voltameter. The intermittent action arises from the quantity of liquid sucked up being in all very small, its flow along the sides of the tube does not begin until the drop formed at the upper part has acquired a certain volume, and the liquid flows in less time than the drop takes to form. A part, besides, is vaporized in proportion to the calorific action of the current.

150. The ascent of the liquid is so rapid, in spite of the resistance presented by the narrowness of the passage, that the small luminous drop may be seen at the upper end of the tube as soon as the lower part touches the liquid.

If the tube be too long for the solution to reach the upper part and fall again outside, the liquid remains at a certain height, which gradually becomes lower as the current of discharge from the batteries becomes weaker. The height of this column varies according to the tension of the current so that it could be used as a gauge of the E.M.F.

The tension of a battery has often been compared with the greater or less height of a column of water, and the quantity of electricity that it supplies with the more or less lavish flow of the liquid, according to the diameter of the outlet. The resemblance is here in a manner materially realised by the mechanical effect of the electric current.

151. Liquid cones.—The effects of suction produced by the electric current may also appear in another form. If a higher tension be used,—that of 800 secondary cells,—and if the electrode be brought near to the surface of the distilled water, the liquid sometimes rises in the form of a *cone* before the spark strikes. This phenomenon, little noticed in static electricity, but nevertheless discovered by Peltier, and which has been for a long time better seen on oil than on water, is much more marked with this kind of electricity, by reason of the greater quantity in play.

If the electrode be formed of a moist pencil of asbestos or filter paper, there appears round the little aqueous cone, which remains suspended at the end of the electrode during the passage of the current, a thick crown of vapour, proceeding from the calorific effect developed by the current in its passage through the liquid.

152. Detonations produced at the extremity of the positive electrode.—If the positive platinum wire be placed in a capillary tube, as in the preceding experiment, but so that it terminates at the extremity of the tube, when it is deeply plunged in the liquid, a shrill noise is heard, and if it be raised after having been two or three seconds immersed, when the end of the tube attains the upper level of the liquid, a detonation is heard similar to that of a fulminating cap. The tube is nevertheless neither broken nor cracked; but the lower opening has become conical and the glass hollowed in the form of a funnel.[1]

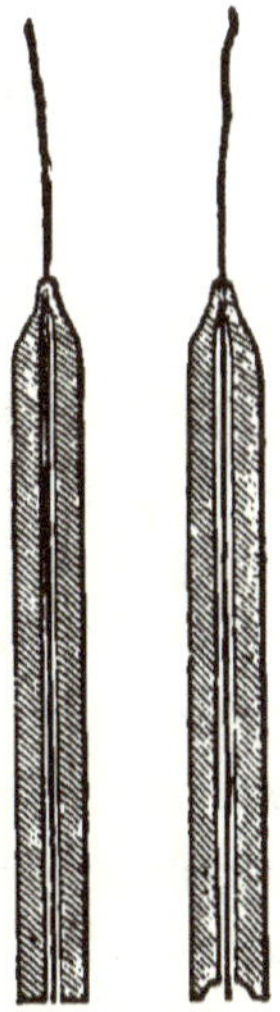

Fig. 56 Fig. 57

---

(1) Comptes rendus, vol. LXXXI, p. 185, July 26th, 1875.

The intensity of this noise is remarkable when the narrowness of the annular space comprised between the platinum wire and the sides of the capillary tube is considered, and if it be also taken into account that this tube be open at both ends; however this phenomenon is more easily produced if the tube be closed at the top. Figure 56 represents the tube before the experiment, and figure 57 the same tube after the detonation has been heard.

153. This phenomenon does not happen, or is much less evident, with the negative wire; for, in this case, all calorific effect of the discharge is carried by the electrode, melts the metal and volatilizes it; whilst, if the wire brought close to the liquid be positive, it reddens without melting, and it is upon the liquid, which itself represents the negative pole (131), that the calorific action of the discharge takes place; vaporization takes place with extraordinary energy round the electrode compressed inside the glass tube, and the shrill noise above mentioned results.

154. We thought at first that the detonation itself, accompanied by pulverization of the glass, was an effect owing to the sudden entrance of air in the tube at the moment when vaporization of the liquid ceased by the interruption of the current. But having since observed that some bubbles of gas appeared at the end of the positive electrode, amidst the liquid vortex caused by condensation of the vapour, and that this gas was formed of an explosive mixture of the elements of the water, consequent upon decomposition at the high temperature produced, and the detonation heard at the moment when the liquid leaves the tube, appears to us to be explained by the ignition of the detonating mixture, which fills the lower end of the capillary tube under the influence of the spark.

This phenomenon would then be allied to those of the same kind observed in voltameters by M. Bertin with much less tension, when the gases produced by the two poles become mixed in one receiver above the electrodes.

There would however be this difference; that the phenomenon here occurs at only one pole in consequence of the extremely high tension of the current.

It is essential, in fact, that the current should have an extremely high tension for the experiment of which we speak to succeed; for two hundred secondary couples are not sufficient; from two to three hundred must be used.

To explain the pulverization of the lower part of the glass tube, an effect that the explosion of a mixture of very small quantities of oxygen and hydrogen barely accounts for, one would be inclined to suspect the formation of some very unstable compound of chlorine which is known to occasion combinations gifted with great explosive power; the more so as we have not been able to observe the effect in question with other liquid than solution of chloride of sodium.

Whatever may be the cause of the phenomenon, it has appeared to us worthy of remark, and we purpose some time to make a special study of it.

155. **Electro-silicious light.**—If the tension of the current used in the preceding experiment be slightly increased without otherwise changing the conditions, by plunging in saline water the glass tube traversed by the positive platinum electrode, a brilliant phenomenon appears at the extremity of this electrode. Not only does the wire melt, but the glass itself is fused in the heart of the liquid, giving forth a dazzling light.

The extremity of the platinum wire which is in the shape of a ball becomes centred in a small mass of melted glass and the light remains very bright during the discharge of the secondary battery,

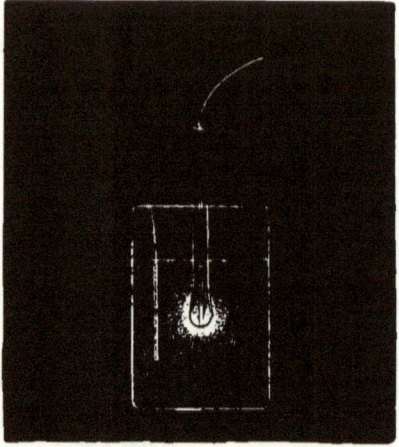

Fig. 58.

until the glass, cooled around the electrode, separates it completely from the liquid (fig. 58).[1]

When using a solution of sea salt in the voltameter, the tension of more than 300 secondary cells is required to produce this luminous effect, but if using a solution of nitrate of potash, the same thing occurs with only 60 or 80 secondary couples.

The manner in which these saline solutions act in presence of silicate of glass, brought to a very high temperature by the electric current, is very varied, by reason of the more or less great fusibility of the silicates formed, as M. Carré has already discovered in mixing different salts with the carbons used for the ordinary electric light.

---

(1) Comptes rendus, vol. LXXXIV, p. 914, April 30th, 1877.

Vitrious light can again be produced by placing the positive or negative electrode against a glass plate, a small distance above the saline solution (fig. 59). It is accompanied by a discharge of white vapour and the glass is at the same time strongly attacked.

This light appears also along the sides of a china basin, as we have already seen (144) (147) figures 51 and 54. It unites its

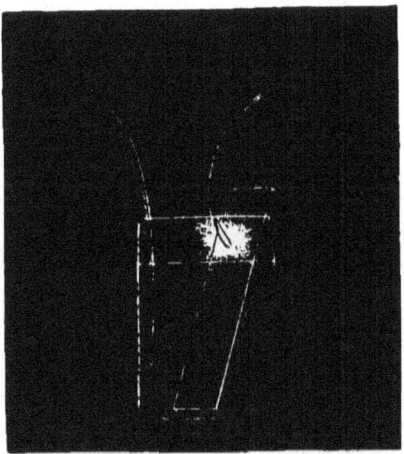

Fig. 59.

brilliancy with that of the voltaic arc in M. Jablochkoff's electric candles when the substance which separates the carbons is a piece of china or kaolin.

The luminous phenomena observed around the glass by means of induction currents by MM. du Moncel, Gassiot, Grove, &c., are also connected with the light in question.

One might be inclined to attribute the brilliancy of this light to the combination of lime, and silica in the glass; but, if the spectrum it gives be examined, it is discovered that there are no perceptible rays, whilst a fragment of calcareous spar placed in the same conditions, though giving forth as bright a light, allows the characteristic rays of calcium to be seen.

156. The rays of silicium being weak, according to M. Kirchhoff's analysis, it is supposed that they do not become visible by reason of the luminous intensity of the spectrum formed. But the silicious origin of this light is proved by the important fact that, in the contact of the electrode with pure silica, it appears in a

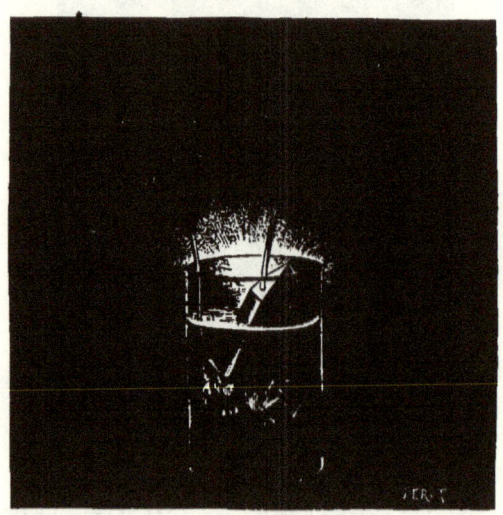

Fig. 60.

state of *quartz hyalin* crystals (fig. 60). For producing it in this case, it is only necessary to use, with the same saline solution, a greater E.M.F. than for glass, about two hundred secondary couples for instance.

The silica itself being necessarily decomposed by these high tension currents, the luminous effect results in all probability from the incandescence of the silica, the remarkable similarity of which to the diamond and plumbago has been shown by M. H. Sainte-Claire Deville and M. M. Wœhler. To distinguish this light from that produced by an electric current between two carbon points, we have given it the name of electro-silicious-light.

157. Crowns, arcs, rays and undulating motions.—1· If the positive electrode of a secondary battery of four hundred couples is placed in contact with the moistened sides of a jar of solution of salt where the negative electrode is already immersed, there may be observed, according to the greater or less distance from the liquid, either a crown formed of luminous particles arranged in a circle round the electrode (fig. 61) or an *arc* bordered with a fringe of brilliant rays (fig. 62), or a *sinuous line endowed with a rapid undulating motion* (fig. 63).[1]

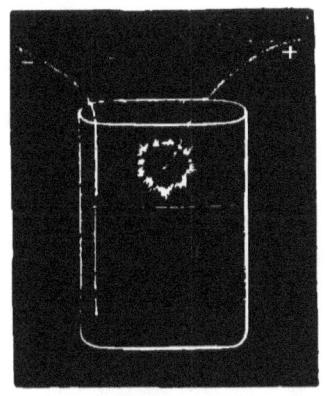

Fig. 61.

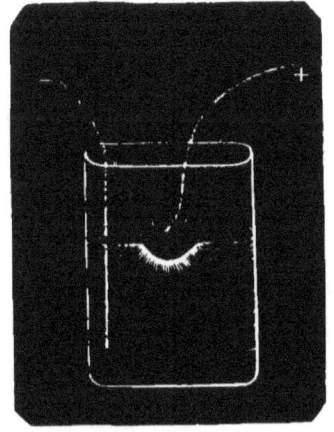

Fig. 62.

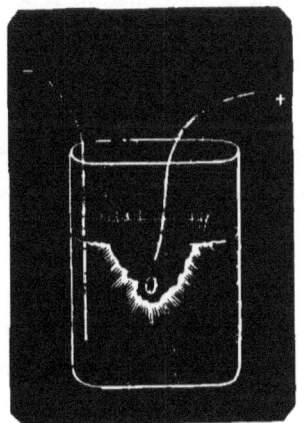

Fig. 63.

---

(1) Comptes rendus, vol. LXXXII, p. 626, March 13th, 1876.

A peculiar noise, continually increasing, is heard, and some vapour escapes from the water in sharp jets, above the rays of fire, as if there were a considerable pressure.

If the wire be still farther dipped in, a luminous closed ring appears; this ring is succeeded by another, and there is thus a creation of brilliant waves in the interior of which the liquid moves with a quick whirling motion.

Sometimes even, round the liquid eddies, little irregular luminous rings are seen to appear, detached from the glass of the electrode.

If the jar used is a tube like a U, which holds but a small quantity of liquid, all these waves end by blending in each other, the liquid becomes entirely luminous and begins to boil violently.

During this time, the deviation of the magnetic needle placed near the circuit, undergoes continual variations.

158. **Electro dynamic whirls.**—The following experiment, which we described in 1860,[1] does not require so great an electric tension as the preceding ones. It may be made with a secondary battery of from ten to twenty couples, or with a battery of from fifteen to twenty Bunsen elements. We will, nevertheless, class it among those relating to effects of electric currents of high tension, because it gives results considerably different from those obtained by employing a much less tension.

The positive electrode is in this case a copper wire, and the liquid in the voltameter is water acidulated to one tenth of sulphuric acid. Whilst in ordinary conditions of the electrolysis of water with this voltameter, by the action of a feeble current, the positive wire becomes covered with a layer of oxide which slowly dissolves

---

[1] Bibl. univ. de Genève, vol. VII, p. 332, April 20th, 1860.

in the liquid (8), if a current of a considerable tension be used, a different phenomenon appears.

The principal oxidation then takes place at the extremity of the wire. A hissing, similar to that produced by red hot metal plunged into cold water, is heard, and the end of the wire gives forth a jet of finely divided oxide, which escapes in abundant flakes and does not dissolve in the liquid (fig. 64).

Fig. 64.

At the same time the wire becomes sharp pointed, and the intensity of the current which traverses the voltameter increases considerably.[1]

If the pole of a magnet and the end of the electrode are brought near to each other, the cloud of oxide moves in an extremely quick gyratory manner, in one direction or the other, according to which pole of the magnet is presented. The rotation takes place according to the laws of Ampère, *in the opposite direction to the hands of a watch, before a north pole (fig. 65), and in the same direction as that of the hands of a watch, when before a south pole (fig. 66).*

The arrows drawn round the whirls indicate the direction of the gyratory movement under the influence of the magnet, and the arrows drawn round the magnet indicate the direction of the

---

(1) The oxide formed in this case appears to be protoxide of copper, rather than oxide, as we have before said (see note to p. 66).

magnetic currents; B is the north pole, A, the south pole.[1]

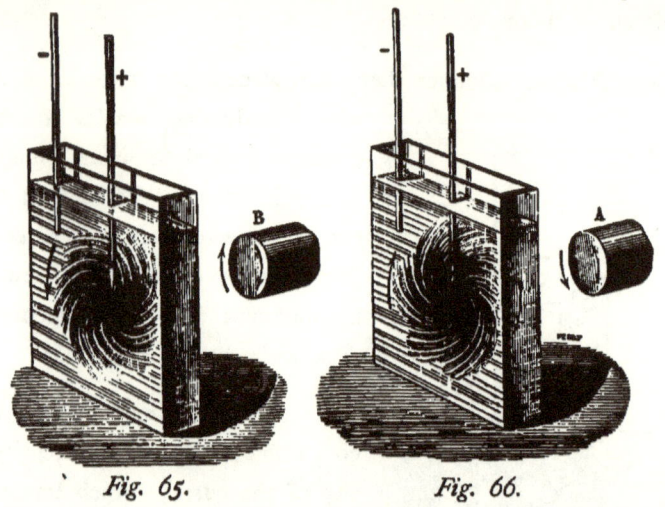

Fig. 65.   Fig. 66.

159. This experiment may be arranged as represented in figure 67. A china or glass basin is placed above an electro magnet and filled with acidulated water; any kind of wire, in

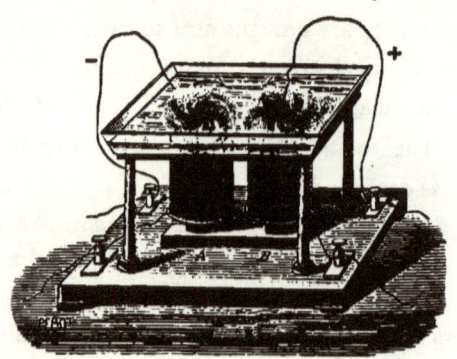

Fig. 67.

connection with the negative pole of a battery composed of

---

[1] This experiment is easily reproduced by projection. We have repeated it in that way at our Lecture at the "Association Polytechnic" in 1861.

fifteen Bunsen elements, is immersed beforehand in the liquid. The copper positive wire held in the hand, is successively plunged into the liquid above each pole of the electro magnet.[1]

The cloud of oxide appears, the whirls are developed, and as the oxide formed does not immediately dissolve in the liquid, but floats in a very finely divided state upon its surface, the two kinds of whirls, in different directions, remain some moments traced on the surface of the liquid after the current has ceased, and even preserve the motion made by the liquid when under the influence of the electro magnet.

Mr. Sylvanus P. Thompson, has since obtained similar whirls by making a magnet, excited by an electric current, act upon iron filings, and has fixed them the same as other magnetic phenomena produced by electro dynamic actions.[2]

The experiment above described may be compared with several others on the rotation of liquids traversed by currents in the neighbourhood of magnets, such as those made by MM. Wartmann, Jamin, etc. But that which more particularly characterises it is the curved spiral form of rotation, in consequence of the magnetic action exercised on the currents radiating round one point formed by the extremity of the electrode, and the distinctness of these whirls is all the greater, as the electrode itself supplies, by its disaggregation, the solid matter necessary to render the motion of the currents visible in the heart of the liquid.

160. **Crater-like Perforations.**—If a sheet of filter paper moistened with salt water, be put in connection with the negative pole of a secondary battery of four hundred elements, and if on‘

---

(1) The same current may at once excite the electro magnet and act upon the voltameter.
(2) See La Nature, p. 179, August 17th, 1878.

the other hand, the moist surface has just been touched with the positive pole, there appears below this wire, with a brilliant light and rise of vapour, a cavity in the form of a *crater*, its edges bristling with innumerable filaments, dried up and entangled one in the other (fig. 68). The positive wire becomes at the same

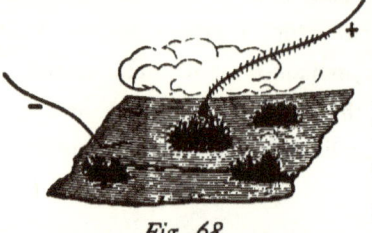
Fig. 68.

time covered with a substance formed by the paste of the paper deposited upon it; thread-like residue also adheres to the electrode over a length of from 10 to 15 centimetres.

The ends of the filaments point towards the positive electrode, so that, if this electrode be placed below the paper, no crater is

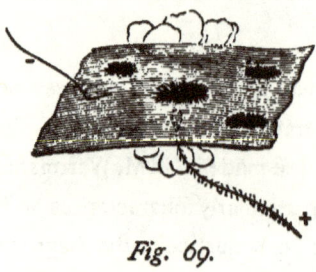
Fig. 69.

seen to project from the upper surface, but only a simple excavation, the stringy ledges of which are as though *sucked in*, and *turned back* towards the point where the positive electricity passes out (fig. 69).

Some filaments, in consequence of their great length and their

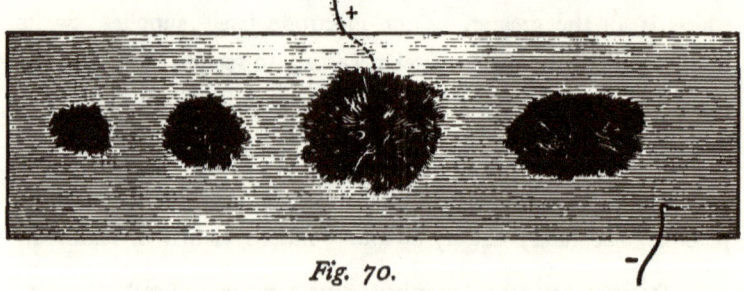
Fig. 70.

instantaneous desiccation, become shaped like a hook at the end.

Figure 70 represents the details of these electric perforations, natural size.

These phenomena are another result of the calorific action exercised by the current, which vaporizes, and instantly dries up the moist fibres of the organic matter, and are also due to its high tension, which produces the effects of attraction or suction, and the mechanical sub-division of the matter subjected to the discharge.

CHAPTER II.

Engraving on glass by electricity.—Other applications.

161. **Engraving on glass by electricity.**—We have previously described (152) an experiment in which a glass tube, with a platinum wire passing through it, serving as electrode to a powerful voltaic current, becomes instantaneously hollowed in the shape of a cone or funnel, in the midst of a voltameter containing a solution of salt. In other experiments (157) on the luminous effects produced by a current of high tension, against the sides of a jar of glass or crystal moistened with a solution of sea salt, we have had occasion to observe that the glass or crystal was strongly attacked at the points touched by the electrode, and that the luminous concentric rings formed all around, remained sometimes engraved on the surface of the glass of the voltameter. We have further discovered that by using, as saline solution, nitrate of potash, a much less powerful electric force was necessary than with chloride of sodium or other salts, in order to produce luminous effects and devitrification.

These observations have led us to apply the electric current to engraving on glass or crystal.[1]

162. The surface of a sheet of glass, or plate of crystal, is covered with a concentrated solution of nitrate of potash, by simply pouring the liquid on the plate, placed horizontally in a shallow basin. Then, in the layer of liquid which covers the glass, and along the edges of the plate, a horizontal platinum wire is immersed, in connection with the poles of a secondary battery of from fifty to sixty elements; then, holding in the hand the other electrode formed of a platinum wire insulated, except at the

Fig. 71.

extremity, the glass covered with the thin layer of saline solution, is touched at the points where the letters or drawing are required to be engraved (fig. 71).

A luminous track appears where ever the electrode touches, and, no matter how quickly they are written or drawn, the characters become distinctly engraved on the glass. If one writes or draws slowly, the lines are deeply marked, their breadth depending on the diameter of the platinum wire serving as electrode; if it be pointed, these characters may be extremely fine.

---

(1) Comptes rendus, t. LXXXV, p. 1232, 1877.

The wire conducting the current becomes thus transformed into a graving tool for glass, requiring no effort in the management of it on the part of the worker, in spite of the hardness of the substance which must be cut; for it suffices to touch the surface of the glass very lightly in order to obtain an ineffaceable engraving.

The corrosive power is supplied both by the calorific and chemical action of the electric current in the presence of the saline solution.[1]

The chemical action of the electric current under these conditions is very powerful, although exercised upon non-conducting matter and simply on its surface; it is even more efficacious in the case of vitreous substances than hydrofluoric acid; for we have thus been able to engrave characters on a plate of Sidot glass[2] which could not be marked by hydrofluoric acid.

Engraving may be done with either electrode; however it requires a current of less strength to engrave with the negative electrode, and the engraving is more distinct.

Although these results have been obtained by using secondary batteries, it is clearly preferable, for continuous work, to use quite another source of electricity of sufficient quantity and tension; either a Bunsen battery of a sufficiently large number of elements, a Gramme machine, or even an alternating current magneto-electric machine.

---

(1) The figures produced on glass by static electricity, and the impressions obtained by M. Grove with inductional electricity, are allied to this corrosion of glass by dynamic electricity. But as the quantity of electricity supplied by electrical machines or induction coils, is relatively very small, and as there is besides no electro-chemical effect, such as that which appears in this case in the presence of a saline solution, these figures and impressions are scarcely visible. To be perceived they require a deposit of moisture, resulting from the breath, which has given them the name of roriḑ figures, since studied by MM. Reiss, Peyré, Wartmann, etc.

(2) This glass is an acid phosphate of lime obtained under special conditions, the discovery of which we owe to M. Sidot, chemical operator at the Lycée Charlemagne.

163. Electric boring or drilling.—We think we ought to point out another application which could be made by the same means, however difficult its realisation may appear at first sight.

We have just seen that, when one of the electrodes which conducts an electric current of considerable tension is brought in contact with glass covered with saline solution, it acts as a graving tool or diamond for tracing lines on the surface of glass and even hollowing it rather deeply.

Rock crystal may be also attacked, in spite of its hardness, by the same method; and, if it cannot be engraved so regularly, at least it breaks into small pieces under the influence of the electrode and finally becomes disintegrated.

Now, in America, they use at present thin black diamonds for cutting hard rocks, and for boring wells or mines.[1]

Could not the use of these diamonds, which are very costly (and gradually waste away by becoming detached from the head to which they are fastened), be replaced by the action of the electric current under similar conditions to those which have just been described, thus obtaining the boring of rocks by electricity?

Platinum electrodes would not be necessary, because it is not in this case the metal of the electrode which is impaired, but the silicious matter when in the presence of a saline solution. Metallic points or projections distributed suitably at the end of the boring bar (a portion of its length insulated and given a rotating movement), would bring the electric current to the surface of the rock which it would be desired to act upon, and would thus replace the numerous black diamonds set or inserted in the end of the

---

(1) See La Nature, August 10th, 1878. Le Sondage au Diamant, (L. Baclé.)

bar, as in the process of diamond boring. The progress recently made in the production of electricity, by mechanical means, would facilitate this application.

164. **Different applications.**—Among the phenomena we have described in the preceding chapter, there are others, such as the electro-silicious light, which might also perhaps be made use of.

If currents of high tension are necessary to discover them, once known, they become easier to reproduce, either with less tension, or with greater tension and less quantity of electricity.

It is thus that a considerable number of these phenomena may be displayed to a certain extent, in a rudimentary state, with strong inductional coils or even with static electricity. Such is the phenomenon of the *sheaf* of aqueous globules, observed with currents of high tension. A conductor in connection with an electric machine or with one of the poles of an induction coil, brought in contact with water, produces a kind of mist, which might be mistaken for vapour, but that the preceding discovery proves to be water reduced to an extreme state of sub-division, or *pulverized* by the electric discharge. Likewise for crater-like perforations (160); by studying with a magnifying glass the holes made by the piercing instruments in static electricity, one finds in them nearly the same characteristics as in those shown more clearly by dynamic electricity of high tension.

Applications of the phenomena above described could then be made with electrical machines or induction coils, when it is not necessary to have at the same time a great quantity of electricity.

FOURTH PART.

Analogy of the effects previously described with Natural Phenomena.—Theories which may be deduced from these phenomena.

CHAPTER I.

Analogy with globular lightning.—On the nature and formation of globular lightning.—Remarks on some cases of globular lightning.— Lightning " en chapelet."

165. The phenomena which we have discovered with electric currents of high tension (135-142) present striking analogies with those of globular lightning and appear to us to be of a nature to facilitate the explanation of this extraordinary form of lightning.[1]

---

[1] "Lightning in the form of a ball, of which we have quoted so many instances, wrote Arago (Notice sur le Tonnere, p. 219), and which is so remarkable, appears to me at present to be one of the most inexplicable phenomena in physics."

"P. 396, *ibid.*—There is but one circumstance in which the physicist does not know how to engender what nature produces so easily; he cannot cause lightning in balls; he does not know how to produce these spherical agglomerations of matter which move slowly without losing the property of striking objects. On this subject there is a blank in science which it would be most important to fill.

We have seen in fact that ponderable matter tends to take a globular form under the influence of a powerful source of dynamic electricity. We first proved this property on liquids and discovered liquid luminous globules (135) fig. 39. By increasing the tension, we obtained, even in the midst of air mixed with vapour from water, real *globules of fire* (138) fig. 42 to 47.

We are then naturally led to think that globular lightning must be produced by a flow of electricity in a dynamic state in which quantity is added to tension.

Thus, it is in great storms, when electricity abounds in the atmosphere, and the discharges constitute a kind of powerful electric current of very high tension, that the lightning appears in a *globular* form instead of taking the simple linear shape similar to that of the sparks of static electrical machines, like that which happens in storms of less intensity.

166. The nature of globe lightning should apparently be the same as that of the globular sparks produced in our experiment.

These globes must be formed, in our opinion, of *rarified incandescent air and gases resulting from the decomposition of vapour from water also in a state of rarefaction and incandescence.*

The water is in fact not only vaporized but decomposed, as has been shown above (154), at the end of one pole, in consequence of the extremely high temperature developed by an electric current of high tension.

167. Although an aqueous surface is not indispensable for forming luminous electric globules, since we have obtained them over a metallic surface (146), the presence of water, or of vapour from water, at least facilitates their formation or tends to give them

more volume because of the presence of the gases furnished by the decomposition of water at a high temperature.

We have more than once observed in our experiments, when all the discharge is devoted to the production of a single phenomenon, electric flames in flattened spherical or cup shaped spherical forms which cover the whole surface of a small jar full of water[1] upon which a current of high tension has just been applied.

Also damp air seems more favorable to the production of globe lightning, and it has often been seen either upon inundated ground after abundant rain[2] or in an atmosphere saturated with moisture. Farther on (186-188) we will quote other examples.

168. We were not considering globular lightning as enclosing a detonating mixture made of gases formed from the decomposition of water, and the noise which often accompanies their appearance as owing to this cause (a noise to which we shall refer further on) (179). These gases are here so rarified that they could not produce an explosion; they are even in an incandescent state and consequently in a condition altogether different from that of an explosive mixture, produced when cold, which would be afterwards suddenly ignited.

---

(1) This jar was about four centimetres in diameter.

(2) See Arago. Notice sur le Tonnerre, p. 46.

At Massa-Carara, September 10th, 1713, during a storm and a deluge of rain, Maffei and the Marquis of Malaspina suddenly saw on the surface of the paved road a very bright flame appear of a partly white partly azure light; this flame seemed strongly agitated but without any progressive motion; it dispersed suddenly after having acquired a great volume.

Ibid., p. 50 (Observation made at Trieste in 1841, and sent by M. Butti to Arago)—"the thunder burst forth at intervals with terrific noise. The street was deserted, for the rain fell in torrents and *the public road was changed into a river*;...the first thing which struck my sight was *a ball of fire which proceeded along the middle of the street*....To give an idea of the size of this fiery globe and of its colour I can but compare it with the moon;..but it had no precise outline; it seemed to be wrapped in an atmosphere of light of which it was impossible to mark the exact limit

169. The formation of globe lightning may be explained in the same way as that of globules of fire obtained in the experiments previously described (137).

The spherical agglomeration of matter subjected to the action of a powerful electric current is, as we have before said, the result of suction or vacuum produced by the passage of the current.

Each of these balls is a kind of *electric egg*, without the glass covering, a voltaic spray (139), which the surrounding medium tends continually to fill; but the abundance of the electric current rarefies the matter as it flows into the electrified centre.[1]

170. The light of these balls, which is sometimes very dazzling as many observers have remarked[2] (188), is explained by the great quantity of electricity in play at the time of their appearance.

The light produced in the electric eggs of the laboratory is feeble because the quantity of electricity of high tension which passes through them is very small. But it is known that in the narrowed parts of the rarefied gas tubes this light is much brighter, and is more brilliant in proportion as the electric machine or induction apparatus employed can supply a greater quantity of electricity.

Among the causes of the brightness of the light sometimes emitted by globular lightning, may be mentioned the incandescense of the cosmical particles of the atmosphere which, though in a very small quantity, add their brilliancy to that of the air and the gases from rarefied and incandescent water vapour.

---

[1] The sprays, sometimes spherical, in static electricity and the sparks observed by M. du Moncel with the induction coil which often terminated in a *ball of red fire* (Notice sur l'appareil d'induction de Ruhmkorff, p. 143) are connected with the same order of phenomena.

[2] Arago. Notice sur le tonnerre, p. 43 and 46.

In fact these cosmical particles contain, besides organic matter, mineral matter such as iron, silica, lime, &c.,[1] substances gifted with great radiating power, at a high temperature. Besides the luminous effects resulting particularly from the incandescence of silica under the action of electricity have been seen before (153). (Electro-silicious light.)

171. The colour of the globes, which is very varied, like that of ordinary flashes of lightning, depends, in our opinion, on the hygrometrical condition of the atmosphere and also on the quantity of electricity in play.

If the water vapour is very abundant, the hydrogen proceeding from its decomposition predominates and the ball inclines to a red colour, as that is the colour most peculiar to rarefied hydrogen, when traversed by a strong current.

If, on the other hand, the electric current is relatively less abundant, rarefaction and decomposition are less complete in its path and the colour inclines towards the violet blue peculiar to rarefied air.

The intervening shades would be explained by the varying proportions between the rarefied gases of the air and of water vapour.

172. The peculiar odour which accompanies the fall of globes of fire, and even of ordinary lightning, may again be explained by the combustion of the cosmical particles joined to that of the matter itself directly struck by the discharge at the point where it reaches the ground. It can be understood that, in the long path of a flash of lightning or in the passage of a column of air by

---

(1) See Les poussière de l'air, by Gaston Tissandier, p. 12, Paris, 1877

the electric current, the cosmical particles encountered would be of sufficiently great number to give forth a perceptible odour by their combustion.

The ozone and nitrous products formed by the combination of the elements of air also doubtless contribute a considerable portion.

173. The noise which accompanies the appearance of globe lightning and which is noticed in the experiment described above (135), proceeds from the rapid vaporization that the electric current developes.

We will add that, during the discharge produced in particular by the positive electrode over the distilled water (140), a very marked sound of blowing is heard, evidently owing to the vaporization of the water which heats under the action of the flame emanating from that electrode much more strongly than under the action of the negative electrode.

174. These considerations, joined to that of the ordinary direction of positive atmospheric electricity, incline us to think that the electric sign of balls which result from the direct flowing of electricity from the clouds should be positive, whilst that of the fires of St. Elma, sprays, luminous columns (189), and other electrical effects produced by induction, must be negative.

175. The gyratory motion sometimes observed in globe lightning would simply result from the reaction due to the flowing of the electric current; the same as in the case of the spiral motion of liquid globules (136), or luminous fibres, composing the ovoid flames (138) formed on the surface of the voltameter.

176. Globular lightening appears either in the shape of a simple fall of balls of fire, more or less numerous, which disappear immediately, or in the shape of a single globe which moves slowly, and remains visible for some time.

In the first case, the globes of fire appear to owe their origin to flashes of lightning of a peculiar kind which we will describe further on by the name of "lightning en chapelet" (188), and the formation of which we will explain.

They are flashes of lightning produced by the flowing of a greater quantity of electricity than that of ordinary lightning and involves the production of spherical agglomerations of rarefied electrified matter in their path.

177. The second case, consisting of the slow passage of a fulminating ball, may be produced in our opinion in two different ways.

We have before pointed out (141), that the globules of fire obtained over water, or even over any kind of conducting surface, by means of an electric current of high tension, naturally followed the movements of the electrode at the extremity of which they appear; so that, if one works in the dark or if the wire serving as electrode and oscillating like a pendulum be concealed by a screen, only a globule of fire is seen moving above the conducting surface.

So, in nature, if a storm cloud charged with a great quantity of electricity happens to pass at a short distance from the ground it may form *a column or water spout of moist air, invisible, and strongly electrified, which serves as electrode,* and produces the flow of the electric current in the shape of a globe of fire which appears at its extremity. This column being essentially mobile, the globe of fire naturally follows all its movements.

178. But the slow motion of these globes may be also produced in another way, although there may be no displacement of a column of moist electrified air.

We have shown in the experiment of the wandering *electric spark* (142) fig. 48, how, in certain conditions, a globular spark can move spontaneously and slowly enough for the successive development of its capricious sinuosities to be seen. It was only necessary to cause the production of a spark of dynamic electricity of high tension between the two armatures of a condenser, the very thin insulating plate of which could easily be pierced where any crack previously appeared. The spark, instead of being of instantaneous duration, wanders along, burning before it the substance even of the condenser and collecting it in a globule of fire.

It may then be admitted that there are formed in the atmosphere, at the point where globular lightning appears, the elements of a condenser in which a layer or column of damp air strongly electrified plays the part of upper armature, the ground the part of lower armature, and the layer of air interposed, that of the insulating plate.[1]

This layer of insulating air being passed by the electric current, the flow appears in a globular form between the ground and the column or damp electrified layer forming the upper armature.

---

[1] The intervention of a strip of insulating air, in the production of the phenomenon has been, besides, already admitted by M. du Moncel: "The slow movement of the ball of fire," says M. du Moncel, "would only be the result of the variations in the direction of this insulating band, or in the current of air which would have occasioned it, variations which would displace the point where the flow of the electric fluid would appear in a luminous state." (Notice sur le tonnerre et les éclairs, p. 51, Paris, 1857).

When the base of this column presents a considerable area, as sometimes happens if it forms a part of the electrified cloud approached very near to the ground, the globe of fire remains in connection with this armature without changing its place, and continues its way alone, passing through the layer of insulating air in an irregular manner, according to the variations in thickness or the resistance presented, just as, in our experiment, the little ball of fire makes its way between the upper and lower armature of the condenser without displacing the electrodes or the armatures.

In the experiment we recall, the spark is, it is true, a globule of solid matter in fusion; but the other examples we have quoted permit us to suppose that spheroids of incandescent gas would present the same phenomena.

The objection may be raised to these comparisons that the natural balls of fire do not appear at the extremity of metal electrodes. But the identity of lightning with sparks from electric machines is now admitted; and although these sparks come from metallic conductors the masses of water vapour which form electrified clouds are considered as similar conductors. A metallic electrode conducting a current of high tension may then be compared with a column of damp air by which the electricity of storm clouds sometimes descends nearly to the surface of the ground.

179. By the preceding considerations it may be explained how the globes sometimes disappear without noise and, in other cases, are accompanied with thunder.

When the thickness of the layer of insulating air which separates the electrified layer of cloud from the surface of the ground becomes too great, in the path of the fulminating globe, and when, on the other hand, the quantity of electricity supplied by the storm cloud does not increase, the electric flow ceases and the globular flame disappears silently, just as the globule of fire produced on the surface of the condenser ceases to appear when the thickness of the insulating plates becomes too great.

180. If, on the contrary, the storm increases in intensity or (the electrified cloud approaching nearer the ground) fresh quantities of electricity arrive on the surface of the layer of insulating air, the flow, instead of continuing to progress in a relatively calm and silent manner in the globular form, strikes sharply like the ordinary discharge accompanied with the noise of thunder.

It can then be understood that, from the point itself where the lightning globe appeared, zigzag or sinuous darts of lightning, which strike the surrounding objects,[1] flash in all directions.

181. But we do not at all mean by that that the noise is due to the explosion of the fulminating ball itself.

If reference be made to the ideas we have given upon their nature, according to the analogies drawn from our experiments, it will be understood that a small mass of air, rarefied and luminous from the passage of the electric current, cannot burst with the noise of thunder and disperse in fiery darts.

The source of the final phenomenon is in the reservoir itself of electricity that the storm cloud encloses and which is

---

(1) Our explanation on the point is nearly identical with that given by M. du Moncel: "The explosion of the ball of fire and the flashes of lightning which it sends forth laterally" said M. du Moncel "would be nothing more than the electric discharge pure and simple, determined by the conducting substances interposed in the band of insulating air in the range of which the meteor would happen to be." (Notice sur le tonnerre et les éclairs, p. 51.)

discharged at the instant when the flow had begun in the form of a globe of fire.[1]

182. When globular lightning appears in the form of a fall of fiery balls which are seen only for an instant (176), the thunder which accompanies this fall should not be attributed to the balls themselves but to the whole of the lightning *en chapelet* from which they are derived (188), and of which they constitute detached grains.

183. The exceptional intensity of the thunder, often mentioned in connection with the fall of globular lightning, is again explained by the quantity of electricity in play, always greater in the manifestation of this phenomenon than in ordinary cases.

The volume of electric fluid, if it may so be called, that is to say, the mass of ponderable matter passed through and rarefied by the discharge, is then greater. Hence, naturally, a greater vacuum is produced.

But how can electricity produce a vacuum? has been asked all this time. Our experiments permit us, we think, to simply reply: by powerful and instantaneous calorific action which developes electricity and vaporizes all matter placed in its path.

The greater part of the phenomena that we have described (135 to 160), is, in fact, only the consequence of the vaporization produced on liquids or humid surfaces by an electric flow in which quantity and tension are both united.

---

[1] We thought at first, like several authors or observers, that the explosion accompanying the disappearance of the fulminating balls, was due to the electricity accumulated in them, but we have modified our ideas on this point since our last experiments. There is doubtless accumulation or agglomeration of electric matter, since there is spherical enlargement of the space rendered luminous by the flow of electricity, but, as we have before stated, it is the electricity pouring from the whole cloud which produces the discharge, accompanied with the noise of thunder, and not the small quantity of electricity in the ball itself.

184. The reason lightning conductors have often proved ineffectual in cases of globular lightning is explained by considering that the appearance of a fulminating ball reveals the commencement of an abundant and continuous flow of electricity from the storm cloud in a particular spot chosen, and that the simple action of influence exercised by the proximity of a lightning conductor would not be able to stop this flow when once begun.

If these slowly moving balls appear harmless in themselves, since observers near to whom they have sometimes passed have received no hurt, they are not the less a source of great danger; for they represent either the extremity of a cloud electrode or the chosen point where it exercises its greatest influence and they forebode an imminent discharge, all the more destructive because of the greater quantity of electricity in play.

The directions in which a lightning conductor can be effective, cannot be too numerous to prevent the formation of this particular point from which an abundant electric discharge may take place.

MM. Melsen's and Perrot's many pointed lightning conductors appear to us specially designed for this object and preferable to the single pointed ones of great height.

A few of the experiments that we have described, such as the spark in form of a flame pointed basket, produced below a positive electrode conducting a current of high tension to the surface of a liquid (143), fig. 50, the perforations produced on damp organic matter with crater-like formation, bordered with filaments turned back and diverging round the point struck by the discharge (160), fig. 68, indicate the form that negative electricity seems to prefer when going to meet, as it were, the positive electricity and neutralizing it.

These experiments are then in favour of the arrangement of a basket of points, adopted by M. Melsen's for the lightning conductors of the l'Hôtel de Ville at Brussels, and strongly support the views of the learned Belgian on this subject.[1]

185. The experiment that we have just referred to (160), fig. 68, besides offering a striking picture of the effects of desiccation produced by lightning upon vegetables and of their division into laths, thongs, or innumerable shoots, explains how they are torn to pieces, or uprooted, and the effects of suction which often accompany discharges of atmospheric electricity.

186. **Instance of globular lightning at Paris, in 1876.** The preceding explanations appear to agree satisfactorily with the facts that we have had opportunity of gathering or of personally observing concerning globular lightning.[2]

The conditions before pointed out (165–167) as being favourable to the manifestation of this phenomenon; that is, the presence of a great quantity of electricity in the atmosphere, constituting by its frequent and continuous discharges a kind of dynamic flow, joined to the production of abundant rain saturating the air with water vapour, were realized on the occasion of two violent storms which visited Paris on July 24th and August 18th, 1876.

It was also proved that lightning fell in several places in a globular form.

On the 24th of July, between half past three and four in the

---

[1] Des Paratonnerres à pointes, à conducteurs et à raccordements terrestres multiples, par Melsens'. Bruxelles, Hayez, 1877.

[2] Comptes rendus, vol. LXXXIII, pp. 321 and 484, July 31st, and August 21st, 1876.—La Nature, 4th and 5th years. September 30th and October 28th, 1876; April 7th, 1877.

afternoon, a deluge of rain mingled with large hail and accompanied with lightning and thunder poured down upon the Place de la Bastille which we were at that moment crossing. There being comparitively little wind, the storm cloud remained nearly stationary for some minutes; the discharges were incessant, and several claps of thunder, following the lightning without any appreciable interval, announced that the latter had struck the earth several times in the neighbourhood.

Making immediate enquiry, we learned that the lightning had just fallen three times running nearly at the same point on the théatre Beaumarchais, in the court and garden of the house No. 28 in the rue des Tournelles, known at the Marais by the name of the hotel de Ninon de Lenclos.

The manager of the theatre, who happened to be in the dress store, a little room situated in the upper part of the building, had seen a shell of fire fall about the size of a hand.

In the rue des Tournelles, a workman living on the fourth storey had seen a globe of fire, about the size of a cannon ball pass along the edge of the roof near to a pot of flowers, only breaking one stem and then falling in the court yard. At the same moment, another workman on the ground floor saw three little balls of fire above the ground of the same court yard, which was at that moment completely flooded.

Over the way, M. Languereau, manufacturer of bronzes, saw in his garden two or three incandescent particles fall, without any clearly defined outline, and drown themselves, to use his expression, in the garden which was transformed into a vast basin by the amount of rain, falling like a veritable water spout.[1]

187. These incandescent particles would not be formed, in our opinion, of ignited matter, but principally of rarefied air and gases from luminous water vapour, like the globes and electric flames produced in our experiments.[2]

However, as hailstones are found having inside them kernels of mineral or organic matter, there may be also found in this kind of luminous particle some cosmic corpuscules borrowed from the atmosphere.

188. **Lightning "en chapelet."**—The storm of August 18th, 1876, was even more remarkable than the preceding one for the intensity of its electric phenomena. It happened after a long period of heat and dry weather and was accompanied by torrents of rain. This storm, the different phases of which we attentively followed from one of the highest points in the neighbourhood of Paris (the heights of Meudon) where we happened to be at that season of the year, gave us the opportunity of observing a very rare kind of lightning, not classed in meteorology, which appeared to us of a nature to throw new light on the formation of globular lightning.

The storm broke towards six in the morning in the neighbourhood of Paris. A vast cloud darkened the sky and gave forth a series of flashes of lightning of great length and very varied shape: some were forked, others presented curves with numerous points

---

(1) The material damages were insignificant, as might be expected from the fall of this column of water, which would easily conduct the greater part of the electric current to the ground. A piece of zinc from the roof of the theatre, lifted up and shot upon the next house, the ignited gas at the end of a lead pipe, and some shocks felt by different persons who witnessed the phenomenon; such were the only accidents heard of.

(2) We have had, in fact, often occasion to observe, that the least cause, such as a breath or draught of air, was sufficient to change the spherical shape, and alter the outline of these flames.

or with unbroken outline. One of them, doubled upon itself, was exactly similar to a curve known by the name of the "*folium* de Descartes."

This lightning appeared to be principally composed of brilliant points similar to the tracks of fire produced on a damp surface by an electric current of high tension.

Towards seven in the morning, at the moment when the storm began to extend over Paris, a very remarkable flash of lightning shot from the cloud to the ground, describing a curve similar to an elongated S, and remained visible for an appreciable instant, forming a chain of brilliant beads scattered all along a very narrow luminous thread (fig. 72).

Fig. 72

This flash appeared to strike Paris in the direction of Vaugirard. The newspapers, in fact, said a thunderbolt had fallen at Vaugirard,

Grenelle, &c., and further, that it had been seen in an ovoid or globular form.[1]

It is probable that the fall of the lightning must have happened simultaneously at different points, and that it divided into several branches or beads when near the ground; for we had only seen one flash reach the earth in this direction.

The rain was very heavy, so that the air through which the discharge passed must have been entirely saturated with water vapour.

The lightning also fell during this storm in a globular form[2] on a house 35, Rue de Lyon. This fact was also mentioned in the newspapers,[3] and we ascertained, by enquiry, that it was true.

---

(1) We extract from newspapers published in Paris, Saturday, August 19th, 1876, the following passages:

"The long expected storm has at last arrived. Towards midnight flashes of lightning began to line the cloud, silently, yet increasing every instant in intensity. Towards four in morning, they followed each other uninterruptedly, like sparks from fireworks.

It was remarked that the thunder claps differed from the ordinary roar. They were not all like the usual crackling, but a series of dull sounds, similar to a cannonade. . .

Lightning has fallen in several places and produced rather curious phenomena.

Thus, on the boulevard de Vaugirard 259, the electric fluid entered by the chimney, crossed a room occupied by a servant, who was fortunately absent, and after having set fire to a bag of linen, left the room, breaking two panes of the window in its way.

... *Nearly at the same instant* a flash struck the house, No. 99, rue d'Assas. The fluid, *in an ovoid form*, destroyed the west gable of the house and shot it to a great distance in the surrounding gardens. Pieces of stone from the cornice, in falling on the balcony of the fifth storey, sent forth thousands of sparks. The zinc which bordered the gable was cut as though with shears.

(2) It is probable, says M. H. de Parville, that the Phenomenon of globe lightning appears more often than one thinks; it may have, until now, escaped the notice of observers not prepared for it. Thus, according to M. Alluard, director of the Observatory of Puy-de-Dome it is not uncommon to see, during a storm, quantities of little balls of fire rebound on the side of the heights.

(3) We also extract this passage: "The storm which yesterday, August 18th, broke over Paris, was accompanied at moments by torrents of rain. At No. 35, Rue du Lyon; lightning appeared in the form of a luminous globe. It also fell in the court-yard of some works, 7, Rue Jules-César (a few steps from 56, Rue de Lyon).

Among other witnesses, a pupil in the chemist's shop situated on the ground floor of this house declares that he saw fall at some yards distant, at the same instant, two balls of fire so bright as to dazzle him, which disappeared on reaching the ground.

Although, from Meudon, we could not see the lightning which struck this point in Paris, on account of the curtain of rain which hid it from view, observation of the lightning "en chapelet" which fell at Vaugirard permits us to think that that of the Rue de Lyon would be of the same kind. In short, those which appeared in the midst of the clouds, as we have already mentioned, presented the appearance of a series of brilliant points rather than of luminous unbroken lines.

189. The quantity of electricity spread through the atmosphere was so great during this storm, that very curious effects of influence, similar to the sprays and St. Elmo fires appearing on the ends of masts of ships, were observed by M. Trécul in the neighbourhood visited by the thunder[1]

190. This kind of lightning appears to us to constitute an indicative phenomenon[2] showing the transition from the ordinary form of lightning, in sinuous or rectilineal lines, to the globular

---

(1) "During the storm which happened on the morning of August 18th," said M. Trécul " between seven and eight o'clock I was writing before my open window. Tremendous claps of thunder which seemed in the immediate neighbourhood were heard repeatedly. At the moment the nearest were heard, or nearly at the same time, small luminous columns fell obliquely on my paper. One of them was about two metres in length—in appearance they resembled lighted gas with an indistinct outline; no detonation took place, only when nearly extinguished, they left the paper with a slight noise." (Comptes rendus, vol. LXXXIII, p. 478, August 21st, 1876).

(2) The form is much more marked and clearly shown in some cases than in others; these special cases are those in which the nature or the form is less hindered or constrained by other influences or confounded with them. We call these cases "striking and indicative cases." Bacon, novum organum, lib. II, § 20.

form. It can be understood that the beads of lightning might acquire a certain volume and become globes of fire.

We concluded from this observation that globes which fall in a greater or less number, accompanied by thunder, and which disappear immediately, could be considered as being derived from lightning "en chapelet."

191. This formation of luminous grains, alternating with darts of fire, would be the result of the passage of electricity across a ponderable medium, and may be compared either with a string of incandescent globules (presented by a long wire melted by a voltaic current, the extremities of which remain for an instant suspended in fusion at the poles of the battery), (99) fig. 23, or to the expansion resulting from the flowing of the whole liquid vein.

Electrified and luminous matter must naturally be dissipated more slowly, in such agglomerations, than the line which unites them, and the longer duration of the lightning may be thus explained.

192. This observation agrees with another of the same kind, quoted by M. du Moncel in the description of a series of flashes of lightning with tracks of long duration. During a storm in London on the night of the 19th of June, 1857, several flashes were remarked "which lasted some moments and did not disappear before being, as it were, melted into beads of light."[1]

193. We have thus been led to propose collecting these examples of lightning of a peculiar character and to classing them under the name of lightning "en chapelet" among the meteorological phenomena.[2]

---

(1) Notice sur le tonnerre et les éclairs par le comte du Moncel, p. 54.
(2) Comptes rendus, vol. LXXXIII, p. 484, August 21st, 1876.

194. Since then, several witnesses have confirmed the fact of the existence of this kind of lightning.

In a communication addressed to the Académie des Sciences, November 20th, 1876, M. E. Renou wrote saying that our observation reminded him of a case quite similar which he had witnessed before.

"During a violent storm which broke out on the evening of July 20th, 1859, at ponts de Braye, commune de Sougé, on the boundary of the departments of Sarthe and Loir-et-Cher, the lightning, said M. E. Renou, seemed to fall on the Italian poplars situated on the banks of the Braye, between 200 and 250 metres from where I was; the lightning made a vertical track, slightly sinuous, formed of balls nearly in contact, exactly like a string of beads and extremely brilliant."(1)

M. Renou also brought forward another argument in support of the explanation we have given of the origin of globe lightning, by comparing the diameter which the grains of bead lightning appeared to him to have at a certain distance, with that of the fulminating globes seen close to by some persons.

"This appearance," said M. Renou "was instantaneous, but, according to the impression which it left me, I estimated the diameter of the balls at the tenth part of the diameter of the sun; an angle of three minutes to 200 or 300 metres would give to these globes a diameter of 0m.20; it is the same diameter which has been given to those globes of fire occasionally seen to pass slowly through the interior of houses without hurting any one present."

195. The R. P. Van Tricht[2] relates how, during a violent storm accompanied by hail, which took place at Namur, July 24th, 1877, and which was considered the worst in that month, one of his colleagues, watching with him, "very distinctly perceived one of those flashes of bead lightning spoken of in the Comptes rendus de l'Académie des sciences at Paris.

196. On the other hand M. Daguin, Professor at the Faculté des sciences de Toulouse, wrote to us lately:

---

(1) Comptes rendus, vol. LXXXIII, p. 1002, November 20th, 1876.
(2) See l'Etude des orage en Belgique, par A. Lancaster. Annuaire de l'Observatoire Royal de Bruxelles, p. 279, 1878.

" In confirmation of the form of lightning in beads, of which you have quoted instances, I can tell you of a flash of that nature which I saw pass from the clouds to the earth. I was then at the Toulouse Observatory and I had classed this case among the strange and varied forms often presented by lightning seen from an elevated position."

197. Other instances of similar flashes of lightning have been more recently published in England:

" On the evening of August 16th, 1877," wrote M. B. Joule (1) "a violent storm took place at Southport.—Among the most brilliant flashes which I observed was one which presented a phenomenon I had never before witnessed. From the point where it left the clouds to its fall into the sea it seemed formed of little detached fragments which gave it the appearance represented below (fig. 73).

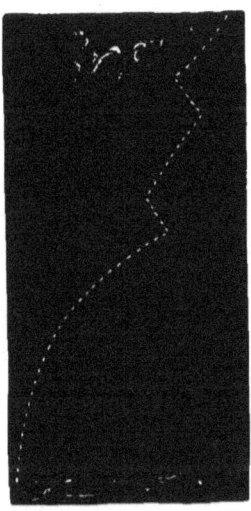

Fig. 73.

198. This latter observation has been supported by another published in the same collection.(2)

Mr. E. J. Lawrence writes—"I can confirm this fact, that lightning sometimes presents a punctuated form. . . . About forty years ago, during a storm, accompanied by heavy rain, which I witnessed at Ampton (Suffolk), flashes of lightning followed each other incessantly

---

(1) On a remarkable flash of lightning; Note read by M. B. St. J, B. Joule at the Society of Physics, &c., Manchester. "Nature," vol. XIII, p. 260, July 4th, 1878.
(2) "Nature" vol. XVIII, p. 278, July 11th, 1878. Remarkable form of lightning by E. J. Lawrence.

for more than half an hour, and about a quarter of them, as far as I can remember, presented this peculiar appearance. Since that time, I have often watched for it, but I have only seen it once more, and then there was but one flash of this kind, among a great number on both occasions, these punctuated flashes were of excessive brilliancy, and presented a curved sinuous form, without sharp angles; one among them had the shape of a nearly perfect fig. 8.

199. The reality of the existence of bead lightning, or simply punctuated (when seen at a greater distance), appears to us demonstrated by the preceding facts, and allows us to form them into a separate class to which we may call the attention of observers.

It would be further interesting, in the case of observing more lightning of this nature, to ascertain if it has been followed by the fall of globular lightning, which might confirm the views we have just expressed.

CHAPTER II.

Comparisons with the phenomenon of Hail.—Upon the formation of Hail.

---

200. Mechanical and calorific effects upon watery substances, or moist surfaces, by discharges of dynamic electricity of high tension (143-145), may possibly throw a new light upon the origin of hail.[1]

The phenomenon of the spray of aqueous globules (figs. 49 and 50) (143), which is projected or thrown forth when a powerful electric current happens to strike the surface of the liquid, shews that a similar effect may be produced when an electrified cloud, or an atmospherical electric current, penetrates another cloudy substance in either its natural state, or less powerfully electrified.

It is true that clouds are not exactly masses of liquid, but those in very elevated regions are composed, we know, of very fine and light ice crystals, the cohesion of which is less than that of ordinary ice, and which may be considered as almost equivalent to a liquid mass in suspension in the atmosphere. It may then

---

[1] Comptes rendus, t. LXXXI, p. 616, et t. LXXXII, p. 314, October 11, 1875, and January 31st, 1876.

be conceived that electrical discharges may, in such a case, result in a similar effect to that which they produce on a liquid, and that the water of these ice crystals, melted and split up at the points where the discharges take place, may be ejected in the form of a spray of globules, as in the experiment.

Moreover, on account of the low temperature of the bulk of the cloud itself, or of the great altitude where the phenomenon takes place, these globules may be instantaneously frozen, and so originate hail stones.[1]

201. According to the greater or less density of these clouds, and the quantity of electricity in play, the calorific or mechanical effects produced by the electric current may alternate, become mixed, or exchanged the ones for the others, the same as was seen in the experiments already described, when the electric current caused either a mechanical effect, such as the expulsion of water in a liquid state, or calorific effect, shewn by an abundant emanation of vapour, according as to whether it met with a quantity of water, or only a moist surface (144) fig. 51.

When the calorific effects dominate in the action of an electric current in the centre of a cloudy mass, the water is no longer only scattered, but vaporized by the current, and this vapour,

---

[1] If the experiment was effected with a much higher tension upon ordinary water, under a receiver, at a very low temperature, the drops emitted would be evidently solidified, and a more complete artifical re-production would be made of the natural phenomenon.

The difficulties in carrying out this experiment being considerable, on account of the space necessary for the enclosed cooled atmosphere, we made an analogous experiment by operating in the ordinary temperature, and taking a concentrated solution of salt (nitrate of potassium), heated nearly to boiling point, so that the drops thrown forth by the electric discharge might be quickly solidified by refrigeration.

The electric current, being brought to the surface of this solution, contained in a cup placed on a pedestal, about two metres from the ground, so as to allow a considerable height for the drops to solidify in falling, the spray is produced, and we thus obtain by electrical means an artificial hail of nitrate of potassium.

immediately condensed in liquid drops at the heart of the cold cloud, might still give rise in this case to hail stones.[1]

202. We are then led to consider hail as arising from the freezing of the finely divided cloud water, vaporized by electric discharges in the lofty and cold regions of the atmosphere.

203. The intensity of the electrical phenomena generally shewn in hail storms, when the flashes of lightning are almost incessant and resemble the continuous discharge of a powerful current of dynamic electricity at a very high tension, shews the importance of the part that must be played by the mechanical and calorific effects which come into action in the production of hail.

On the occasion of violent hailstorms which took place in Switzerland and France, on the 7th and 8th, July, 1875, there were from eight to ten thousand flashes of lightning per hour, resembling an immense conflagration.[2]

An idea may be given of the enormous quantity of heat and water vapour that can be produced by such a torrent of electricity

---

(1) It seems at first sight that a substance so finely divided as water vapour could only produce extremely small hailstones. But it must be taken into consideration that, before congealing, the vapour necessarily passes through condensation, and that if it is produced rapidly and in great abundance, it may be quickly condensed in drops of considerable size against the cold parts of the cloud, just as it is condensed in drops against the interior surface of the lid of a vessel full of water at boiling point.

(2) Upon two hailstorms, &c., by M. D. Colladon (Comptes rendus, t. LXXXI, pp. 104, 446 et 480). "The electric phenomena," writes M. Colladon, "were very remarkable about the central portion of the hail cloud; the flashes of lightning followed each other with such rapidity, between midnight and one o'clock, that we counted on the average two to three flashes per second, which would make eight to ten thousand per hour,—at every place visited by this storm the lightning has been compared with a gigantic conflagration, so permanent did the glare appear."

During a violent hailstorm which took place on the 25th July, 1877, in the High Alps,— "the lightning lit up the sky with an uninterrupted light; the thunder claps followed each other without interval." Note de M. Le Capian. Bulletin de l'Association scientifique de France, p. 367, September 9th, 1877.

in the heart of the clouds when we see the amount of vapour liberated in the experiments given above.

204. The violent oscillations produced in the middle of clouds from which hail is falling, and the rapid transformation of the cirrhus into nimbus also support this view; for the suddenly appearing nimbus can only arise from rapid vaporization, and from water condensed from a portion of the cirrhus.

The multiplication of sifts in hail clouds, and their distorted form, are likewise explained by the effect of electrical discharges, if we compare them with the effects produced by high tension currents upon moist substances (160).

205. The high wind, which nearly always accompanies hail storms, may be attributed to the rarefied air caused by the electric current quickly vaporizing the moisture which it meets with in its passage, and to the inflowing of the surrounding air, which immediately fills up the vacuum formed.

206. The noise which precedes or accompanies the fall of hail is due to the electric current penetrating the cloud, and consequent disintegration or vaporization resulting, just as the noise produced by the passage of a high tension current in a liquid or upon a moist surface, is due to the projection in the form of a spray of the water globules, or to the rapid issue of jets of steam (144 to 157).

207. The lightning which accompanies hail storms, with or without thunder, arises from the fact that, in this collision between two moist masses possessing great mobility of form, sometimes the one penetrates more or less deeply into the other, just as, in the action of a high tension current upon the surface of a liquid,

the emanation is produced in the form of luminons rays, accompanied with a slight noise, when the liquid is rendered strongly negative by the electrode being completely submerged, whilst loud sparks appear if the liquid be, on the contrary, strongly positive, and the discharge takes place at the negative electrode (135 to 136).

208. In the same way it may also be explained how hail can be produced without apparent electrical manifestations and yet owe its origin to the presence of electricity.

We have, in fact, obtained in our experiments the production of vapour, even without luminous phenomena, when the quantity of electricity given by the high tension current is very small.

In the same way, without any visible lightning or perceptible noise of thunder, there may be a production of vapour in the atmosphere, and solidification in the cold regions, in the form of small hail when there is only a small quantity of electricity.

209. The length of time occupied by the fall of hail at any one point being sometimes very short, may be explained by the brief duration of the electric discharges,[1] and by the strong wind which accompanies the storm cloud, and rapidly drives it towards other points.

210. The fall of hail in paths sometimes so narrow that in the same place and in the same quarter of a town points only a very short distance apart shew no sign of it when the portion between is alone affected,[2] may be explained by the vaporization and

---

(1) We have been able to point out at Paris, where hail storms have generally but a very short duration and a relatively low intensity, falls of hail which only lasted from half a minute to a minute, and followed almost exactly the appearance of lightning and claps of thunder (28th March, and 3rd and 25th May, 1876).

(2) This was also pointed out at Paris in the course of the year 1876.

solidification of the water around the tracks made by the lightning, always much longer than they are wide.

As to the long and wide belts of hail which cover a great extent of country, these may naturally be caused by the passage of the cloud itself under the action of the wind accompanying it.

The width of the belt corresponds with that of the cloud or group of clouds, and its length with the distance traversed.

211. The tracks of rain comprised between two belts of hail may result from the interior mass of the cold clouds, through which the discharges take place, being re-warmed by the frequency of the flashes of lightning, the scattered or vaporized water is only condensed in the form of rain in the intermediate part, whilst solidification may still occur on the borders and continue throughout the whole of the passage.

212. The intermittant periods and returning force of the storm which may be remarked either in the fall of the hail or in the gusts of wind which accompany it, are quite analagous to those noticed in our experiments when the electric current debouches upon a moist surface (145), and may be explained in the same manner.

When the electrified cloud has reduced in the form of vapour a portion of the cirrhus into which it penetrates, there is a moment before it can meet with a new substance to vaporize; but the remainder of the cirrhus immediately fills the vacuum formed; a fresh discharge is produced, and, consequently, another emission of vapour, or of finely divided water, and formation of more hail stones.

213. The ovide or pointed form of the hail stones, and their sharp corners or protuberances, may be attributed to their electric origin; for in the experiment of the "spray" (143 to 144), the globules have also an ovide form, and the spark whence they are projected has the appearance of a crown of flaming points.

In other experiments, where the current acts upon a moist paste, which preserves the shape arising from the action undergone, craters may be observed bordered with many pointed protuberances exactly characteristic of the passage of the electric current (160).

214. The light sometimes emitted by hail stones is very likely due to electricity. Although, in our experiments, we could not distinguish whether the aqueous globules have their own light or partake of the reflection of the spark, it is probable that the electric flow imparts to them a brief phosphorescence, since, with a greater E.M.F., the moist air itself becomes incandescent (138).

215 The interior structure of hail stones varies, as we know; some present a radiating structure and seem to have been formed from a single jet, others have an opaque white centre, covered with layers of ice, alternately opaque and transparent.

The formation of the former may be explained, as we have already described (200), by the production of an electric spray of aqueous globules, immediately solidified in the volume they possess at the moment in which they are emitted, or by the water vapour produced under the calorific action of the discharges, condensed in large drops and immediately frozen within the circle of the cloud at a low temperature (201).

In our experiments, the greater the quantity of electricity given by the high tension current, the greater are the globules emitted so, in nature, the largest hail stones are produced in the storms where the electrical manifestations shew the greatest intensity.[1]

216. The structure of the hail stones formed of alternately opaque and transparent layers seems to prove a progressive development within the body of the clouds.

This increase of volume has been attributed to various causes, all of which may be taken into consideration:—either to an oscillating motion in the hailstones, as supposed by Volta, when the arrangement of clouds was favourable;—or to their fall across a great thickness of clouds, as presumed by M. l'Abbé Raillard[2];—or, according to MM. Saigey, Daguin, Fron, Faye, Secchi, &c., to their prolonged spiral motion under the influence of the whirlwinds which usually accompany hail storms, and which have been observed by Lecoq and by MM. de Tastes, Severtzow, Buchwalder, &c.

These whirlwinds may be able to suspend the hail stones for a certain time in the clouds, and so help to increase their volume.

But, in order to explain the alternately opaque and transparent layers, we thought that might be accounted for by successive vaporizations and solidifications, added to the spiral motion.[3]

The opaqueness of the cloudy centre which forms these hail stones, seems, in fact, to prove the sudden chill and solidification

---

(1) In the storms in Switzerland, already quoted, "the size of the hailstones," writes M. E. Plantamour, "attained proportions unusual in our latitudes, and a strong west wind transformed them into veritable projectiles, breaking everything in their passage; some of the size of a walnut, a pigeon's egg, and even a hen's egg were not rare. The following day some were picked up six centimetres across their widest dimension, there were even some nearly one decimetre . . . . . During the fall, there were several flashes of lightning per second, the sky seemed set on fire." (Bibl. univ. de Genève, t. LIX, page 339. 1887.)

(2) *Annuaire* de la Société météorologique, p. 129. 1865.

(3) *Bulletin* de l'Association scientifique de France, p. 49. October 31st, 1875.

of water vapour; for we know that it is the characteristic of rapid crystalizations to lead to distorted and not transparent crystals.

This first core being formed, the spiral motion in the body of cloudy moisture produces all round it a layer of ice, more slowly formed and, consequently, transparent. After another electrical discharge a fresh emission of vapour takes place, and, at the same time, new hail stones appear; those which are still turning may be again covered with a second layer of vapour which is quickly frozen into a snowy condition, and so on.[1]

217. As regards the origin of the formation of these whirlwinds of hail, the experiments above described under the name of electro-dynamic whirls (158) (Figs. 65 and 66), and which we shall refer to further on (221), in order to explain the spiral motion of waterspouts and cyclones, have led us to attribute it to the magnetic influence of the earth. The appearance of hail is, in fact, connected, as we have just seen, with the presence of great quantities of electricity in the clouds, the discharges of which constitute veritable electric currents, of short duration, it is true, or rather intermittant, but possessing all the characteristics of a powerful current of dynamic electricity.

Electric currents may turn under the influence of magnetic action, and with so much the greater rapidity as they pass through more mobile conductors. When these currents radiate in all directions at the centre of a liquid, as in the experiment in question (158), the spiral motion produced under the influence of

---

[1] However, one may demur by saying that the R. P. Sanno-Solaro has artificially obtained small masses of frozen water all at once, without the successive addition of fresh quantities of water, and still shewing concentric layers alternately opaque and transparent.
It would appear from this that the whirling movement is not absolutely necessary in order to account for this kind of formation of hail stones. But in that case their formation may be simply referred to the preceding explanation, which we have already set forth (215).

a magnet, and made visible by a cloud of semi-metallic powder detached from the electrode, works in a spiral form with an extraordinary rapidity.

Columns of cloudy or moist atmosphere, powerfully electrified and changeable in all directions, should then likewise take a rapid spiral motion under the influence of the earth's magnetism, and so raising into whirlwinds the hail which accompanies the discharges.

218. Thus, electricity appears to be connected with the production of hail by the variety of its effects, either mechanical, calorific, or magneto-dynamic.

The part played by the winds is, no doubt, important; they entrain, divide, or re-unite in their passage the electrified cloudy masses; they bring together those strongly charged with electricity with those less charged; they raise them towards the cold regions of the atmosphere, or facilitate the fall of temperature around them necessary to solidification; they also direct them, according to the configuration of the ground, towards the points where we notice that hail seems to appear by preference.

But these are the accompanying causes which only prepare conditions favourable to the production of hail, whilst electricity is, in our opinion, the actual cause which, by its presence in the clouds, and the instantaneous power of its discharges, causes the sudden formation and fall of this phenomenon.[1]

---

[1] Comptes rendus, t. LXXXII, p. 316, January 31st, 1876.

CHAPTER III.

Comparisons with Waterspouts.—Upon the formation of Waterspouts and Cyclones.

219. The effects obtained by means of high tension currents of electricity, which have been described above (146 to 151), present a great similarity to those of water-spouts, and shew the importance of the part that must be played by electricity in these great natural phenomena.

The experiment shewn (fig. 52, page 129), in which a liquid vein or column, powerfully electrified, flows above a magnet, reproduces the principal effects of water-spouts, in infinitely reduced proportions, it is true, but with their most characteristic features;—the noise they make, the mist formed around them, the luminous traits or silent lightning with which they are streaked, the fiery globes which sometimes appear at their extremity,[1] and the boiling of the water when they touch the surface of the sea;

---

[1] We may, in fact, notice, in this experiment, at the point where the column meets the surface of the liquid and all round the spark, small luminous liquid globules at the same time that the steam and finely divided water appear.

so that these phenomena may be compared with positive electrodes of liquid or vapour, from which powerful electric currents from storm clouds escape toward the earth or the sea.

220. This experiment also leads us to attribute the spiral motion of waterspouts[1] to the flow of the electric current under the influence of the earth's magnetism; for this motion takes place in precisely the same manner as in the experiment in question; that is to say in the opposite direction to the hands of a watch, supposing the observer to be placed in the northern hemisphere, and in the same direction if the observer be placed in the southern hemisphere (146).

221. If it be considered that this direction is the same as that taken by cyclones, that the rotation of these great atmospheric currents acts in a spiral form, according to the observations made by numerous sailors,[2] like the electro dynamic spiral motions already observed when an electric current, debouching at a certain point, radiates in all directions over a magnet (158 to 159); if it be also remarked that these spiral motions are accompanied by the most intense electrical manifestations at their starting point in the tropical regions,[3] and that cyclones seem to be developed round a point called the "eye of the cyclone," which is a veritable

---

(1) Most Waterspouts, either on earth or sea, are accompanied, as we know, with a spiral motion. At the approach of a water-spout "the surface of the sea begins to be disturbed; we see the water foam and turn slowly, until the rotatory motion becomes accelerated." . . . (Dampier, Voyage autour du Monde.)

(2) V. Notes sur la forme des cyclones dans l'océan Indien, etc., par Meldrum, directeur de l'observatoire de Maurice. According to Mr. Wilson, manager of the meteorological Observatory at Calcutta, the form of cyclones in the Gulf of Bengal is also rather spiral than circular.

(3) See the works of Reid, Piddington, de la Poey, de la Havane (Tempêtes électriques. *Annuaire* de la Société météorologique, 1855) de Marié-Davy (Les mouvements de l'atmosphère et des mers), de A. Le Gras, Bridet, Roux, Zurcher et Margollé (Trombes et cyclones), etc,

furnace of electricity, we may be permitted, we think, to attribute these formidable phenomena to the magneto-dynamic rotation of atmospherical electric currents, to which the clouds act as moveable conductors, and the movement of which is communicated to the volume of the air surrounding them.[1]

222. We will add in support of these assumptions, as we have already done with regard to whirlwinds of hail, which have, we believe, the same origin (217), that the speed with which these electro-dynamic motions are produced in our experiments, is very great,—the electric current not being confined to metallic conductors, but being free to expand from a single point in all directions through the body of the liquid.

Looking at the rapidity of these motions, we can conceive the power that these atmospherical currents may acquire when free to move in every direction, and charged with a great quantity of electricity radiating equally in all directions in the body of the atmosphere, and transformed by the earth's magnetism into spiral motions.[2]

223. The spiral motion of simoons, as frequently seen in India, the electric origin of which has been demonstrated by

---

[1] Although the spiral motion of cyclones may have been generaly attributed, by most of the writers upon this question, to the meeting of winds of contrary direction, or of different rates of progress, we will add, however, that Reid himself, one of the authors of the *Lois des Tempêtes*, assumed that "electro-magnetism had perhaps some connection with the rotary character of storms, and their opposite motions in the different hemispheres." "(H. Piddington, *Guide du Marin sur la loi des tempêtes*, p. 171, Paris, 1859) . . . . but all this is so speculative," adds Piddington, "that we will limit ourselves to merely pointing it out."

[2] We will not here consider, be it understood, the motion of translation or trajectory of cyclones, which depends upon the direction of the higher and regular winds, combined with the earth's motion of rotation.

Dr. Baddeley, may likewise be explained by the influence of the earth's magnetism.[1]

224. The experiment above described (146), further proves that waterspouts, even when they are not accompanied by any sign of electricity, may, nevertheless, be electrically charged, and owe their spiral motion to the actual presence of this electricity. The fact is, they form, in this case, a conductor sufficiently perfect for the electric current to flow through it, without being transformed into heat or light.

225. The same experiment also proves that waterspouts must be charged with positive electricity; for if they were negative, the spiral motion would take place in the opposite direction to that now observed in either hemisphere.

226. The formation of waterspouts, or the descent of these cloudy appendages towards the earth, has been attributed, with some reason, by Brissou[2] and Peltier,[3] to an electrostatic attraction between the clouds and the earth. To this very naturally attractive force may be added a carrying action, like many

---

(1) The lighting itself sometimes presents a spiral form. Coulvier-Gravier (Recherches sur les météores), M. de Fonveille (Eclairs et tonnerre), and other observers have quoted various examples.

This kind of lightning arises, according to all probability, from the same cause, and seems to us to be comparable with a neat electro-dynamic experiment, due to M. Le Roux (Annales de Chimie et de Physique, 3e série, t. LIX, p. 409, 1860;—et Traité de Physique, par Daguin, 3e édit., t. III. p. 649), in which a very fine and very flexible wire, suspended vertically, raised to a red heat by an electric current passing through it, becomes spontaneously curled in a spiral form, with great rapidity, round a magnet.

The luminous electric flow of the lightning must also turn under the magnetic influence of the earth, and, assuming that this is a descending current of positive electricity, the course described would be a right hand spiral in the northern hemisphere, and left hand in the southern hemisphere.

"See for more detail upon this subject a note published in La Nature, April 7th, 1877, p. 300."

(2) Brisson. Traité de Physique, t. III, p. 418, Paris 1803.

(3) Observations et Recherches Expérimentales sur les causes qui concourent à la formation des trombes, par Ath. Peltier, Paris, 1840.

examples afforded by dynamic electricity, which tends to facilitate the flow of water from the electrified cloud. It can easily be conceived that a very dense mass of clouds, powerfully charged with electricity, might give rise to the fall of a watery column when it passes sufficiently near the earth or the sea.

227. The tidal wave, which often accompanies cyclones,[1] the "bores" (seiches) of the Swiss lakes, consisting in a sudden rise of the waters in the form of waves or undulations, especially at the narrowings of the lakes, which are particularly produced during violent storms,[2] may be also explained by electric actions, as Bertrand and other observers have thought.

The experiment we described under the name of "electric bore" (147), figs. 53 and 54, in which a high tension electric current gives rise to one or more small waves upon the borders of the surface of the liquid considerably raised above its ordinary level, shews that a current of atmospheric electricity may depress or elevate liquid masses, just like a breath or strong wind, and goes further to prove the electric origin of these phenomena.

228. The phenomenon of the ascension of a liquid column, produced by the actual flow of a powerful electric current, which we described above under the name of voltaic pump (148), fig. 55, the aqueous cones formed below an electrode which conducts the current to the surface of the liquid (151) allows the explanation

---

(1) This coincidence of the tidal wave with the cyclone is very remarkable; there is no instance of such a storm having struck La Réunion without being preceded by a phenomenon of this kind." (J. Rambosson, Histoire des météores, p. 243).

Piddington quotes a case observed at Ramsgate, "from the sudden rise and fall of the tidal wave at this port in August, 1846; they occurred three at a time by unequal undulations during a heavy storm and exactly at the same time as a strong discharge of what is called the return shock of the electric fluid." (Guide du marin, p. 136).

(2) See the recent works upon this subject by M. Forel, Bibl: univ. de Genève, août et September, 1878.

of the effects of very strong suction produced by waterspouts, and of conceiving especially how, in waterspouts of tubular appearance, this suction, acting throughout the whole length of an electrified column, can raise water to an indefinite height, a fact which has caused these phenomena to be called by the name of "pumps" or "syphons" in certain parts of the world. The water sucked up may arise from the sides of the vapourous channel itself, and thus is explained the fact of the absence of salt in the water falling back from these marine waterspouts.

229. The phenomena produced by static electricity also shew, in a more feeble degree, effects of suction and evaporation, of which Brisson and Peltier have already pointed out the similarity to waterspouts. But these that we have observed with strong currents of dynamic electricity, combining at the same time both quantity and tension, appear to admit of still closer comparison with the conditions of nature, and we think we may conclude from this experimental study, that waterspouts and cyclones are powerful electro-dynamic effects produced by the combined forces of atmospheric electricity and the earth's magnetism.[1]

---

(1) Comptes rendus, t. LXXXI, pp. 185 et 616. July 26th and October 11th, 1875;— et t. LXXXII, p. 220. January 17th, 1876.

CHAPTER IV.

Comparisons with the Polar Aurora.— Upon the formation of the Aurora, and the origin of Atmospheric Electricity.

---

230. The well known experiment of De la Rive upon the rotation of electric gleams, produced in a vacuum round a magnet[1] has already suggested the electric origin of the polar aurora, and its connection with the earth's magnetism.

But a certain number of circumstances which accompany their appearance still remain to be explained.

The experiments above described (157), figs. 61 to 63, in which the electric current comes into the presence of watery substances or moist surfaces, as in the atmosphere, appeared to us to present a series of phenomena exactly similar to those of the aurora.[2]

231. In fact, we recognise here, in spite of the diminutive proportions, crowns, luminous arcs with brilliant fringes of rays, regular or twisted, and possessed of a rapid undulating motion.

---

(1) V. Traité d'électricité, par A. de la Rive, t. II, p. 248, et t. III, p. 289.
(2) Comptes rendus, t. LXXXI, p. 185, July 26th, 1875; et t. LXXXII, p. 626, March 13th, 1876.

This undulating motion, especially, offers a perfect analogy with that compared in the aurora to the undulating movements of a serpent, or to the folds of a cloth shaken by the wind.

232. Although the yellow light prevails in these experiments, on account of the salt solution used, there may also be observed at points where the water arising from the condensed steam is less charged with salt, purple and violet tints similar to those of the aurora.[1]

233. The rays from the luminous arc of the aurora should arise from the penetration of the electric current into the moist or frozen masses with which it meets, in the same way as those observed in these experiments. The resulting vacuum being gradually re-filled, fresh rays keep continually forming, and thus is explained how the points of light from the aurora scintillate, or appear shot forth and renewed every moment.

234. The dark circle or segment formed in the aurora by the mist or cloudy veil, with which the electric current comes in contact, corresponds in the experiment with the moist circle or segment surrounding the electrode and round which the voltaic current expands.

The portions in the immediate neighbourhood of the point whence the electric current flows, being vaporized, it is only at a certain distance that the electric wave is arrested and transformed into heat and light.

235. The similarity of form between the luminous arc produced in our experiments and that of the aurora is also very

---

[1] See the descriptions of the polar aurora in the works of A. de Humboldt (Tableaux de la Nature, Cosmos), Bravais, Lottin, &c. (Voyages et Scandinavie), Arago (Notices Scientifiques), Piazzi-Smyth (Observations made at the Edinburgh Observatory, 1877, vol. XIV, pl. 5, 6 et 7), Liais (Les Mondes, t. XXXVIII), etc.

striking. The current, passing in this manner through the voltameter, arises from the solution not entirely surrounding the electrode. But if the electrode be more deeply immersed, waves or complete luminous circles are produced (157), the same as in the aurora, the arc of which is often considered as the portion visible to an observer, of the complete luminous circle.

236. In the same experiments, we have noticed that the liquid is violently disturbed by the electric current; whirlpools are formed by the meeting of the electrified waves with each other, and, if we only use a small quantity of liquid, a luminous boiling, corresponding to that fluctuation in the light which also characterises the polar aurora, takes place.

237. As the electrode penetrates deeper into the liquid, the water vapour is set free in proportionately greater abundance. This phenomenon, of which the strongest batteries of static electricity hardly allow us to form a suspicion, is an important matter for consideration; for it explains, in a natural manner, the heavy falls of rain or snow which are always noticed during the continuance of the aurora.[1]

238. The production of strong winds consequent upon the appearance of the aurora-borealis, as we have already remarked with respect to hail, shews that the discharge of a great quantity of electricity in the atmosphere causes the formation of powerful atmospheric currents by its calorific action, and the instantaneous vaporization and rapid condensation resulting.

---

[1] "Often the aurora-borealis is accompanied with hoar frost, and the greater part of them are followed by heavy falls of snow or rain, or by violent gusts of wind and storms." (Extrait d'une communication de Mecker de Saussure à Arago. Notices scientifiques, t. I, p. 694.)

239. The noise often heard during the aurora is due, like that in our experiments, to the vaporization produced in the track of the electric spark penetrating a moist substance.

"This noise is," they say, "especially intense when the rays dart forth with rapidity."[1] The noise in the voltameter is also so much the more intense in proportion to the length of the rays fringing the luminous arc, and the rapidity with which they are formed in the body of the liquid.

240. The magnetic disturbances caused by the aurora may be re-produced in these experiments by placing a magnetized needle close to the circuit. The deviation increases or diminishes according as the luminous arc is more or less developed in the liquid.

241. From these facts we may further conclude that the aurora must be produced by a flow of positive electricity; for the luminous phenomena are similar to those of the positive electrode in the voltameter, and the negative electrode shews nothing like it.

242. But is the polar aurora caused by a discharge between the positive electricity in the atmosphere and the supposed negative electricity from the earth? If that were the case, thunder should be very frequent at the poles, or the gleams and luminous sprays upon the projecting points of the earth, forming the counterpart of the phenomenon passing in the air, would be observed. Now,

---

(1) Kaemtz. Traité de météorologie; traduction de Ch. Martins, p. 428.

The existence of such a noise has been doubted by some observers; but the mass of evidence from people inhabiting the region of the aurora-borealis goes to prove that this noise is sometimes heard, no doubt when the height at which the aurora is produced is not too great. (V. Arago. Notices scientifiques, t. I, p. 693.)

This is what Dr. Hjaltalin writes in a memoir upon the aurora-borealis: "I first directed my attention to discover if any noise did or did not accompany the aurora-borealis; I feel sure that this noise does exist, although it is comparatively rarely heard, and I have only heard it six times in a hundred observations." (V. l'Année scientifique, 1864, par Louis Figuier, p. 107.)

actual observation shews that it is not so. We are then led to think that it is the partial vacuum in the high regions which, acting like an immense conducting envelope, plays the part of the negative electrode in the experiments referred to above, and that the positive electricity flows towards the planetary spaces, and not towards the earth, across the mists or frozen clouds which float above the poles.

243. Regarding the origin of this polar electricity, it was admitted that it came from the equator and tropical regions. But it may be objected that the electrified clouds must become discharged during so long a passage, and, in fact, we know that storms are more and more rare as the poles are approached. Some analogies deduced from our experiments, which we shall find further on (Chap. VI.), having led us to consider the heavenly bodies as charged with positive electricity,—the only kind of electricity which, perhaps, does exist,—we might be inclined to regard the earth itself as charged with positive electricity, liberated from the ground and the sea by means of simple *emission*, and radiating from the whole surface at the poles as at the equator, producing very different effects in the atmosphere, in consequence of the entirely opposite meteorological conditions prevailing in these regions.

244. The positive electricity thus arising from the earth would not appear to be, in our opinion, the result of production or generation, properly so called, by physical or chemical causes. It could not be due to evaporation, nor to friction, nor to thermo-electric action, but should arise from a primary charge or store of electricity proper to the earth itself, carried by it from the origin of its formation and which would tend to be wasted, in the same way as the heat it possesses, with extreme slowness, on account of the large mass.

245. This electricity, penetrating the atmosphere, would consecutively attain the higher strata in which the air, becoming more and more rarefied, affords immense conducting spaces, and would spread from thence into the planetary regions. The lower beds of the atmosphere, near to the earth, not being rarefied, we may conceive that the positive electricity may only appear in a very small quantity and accumulate in the greater altitudes. Thus might be explained the increase in quantity of positive electricity in proportion as we rise in the atmosphere.

246. The earth would not then act, in our opinion, as opposed to the atmosphere, like a body which produces electricity by the friction of another body taking up the contrary electricity. Because we should then be led to admit that this production takes place at the point of separation between the earth and the atmosphere, without any apparent physical or mechanical action other than the slow evaporation at the surface of the sea. Now, it is known at the present time that the phenomenon of evaporation is not in itself a source of electricity.

Vápour formed above the sea appears to us to merely constitute a prolonged conductor in the atmosphere of the liquid conductive mass of the earth, from which positive electricity emanates in consequence of its high tension. It may thus be explained how the liberation of electricity may be greater over the sea than over the solid crust of the earth.

Water vapour facilitates the diffusion of electricity in the air, which, at its ordinary pressure, acts like an insulating body.

247. We may also comprehend, according to this hypothesis of the earth considered as an electrified body throughout its

entirety, that this electricity itself might be set free by means of eruption, and form volcanic storms, always accompanied by lightning and thunder, and happening simultaneously with earthquakes, which are connected with the internal movements of a liquid melted mass, charged with electricity. The vapour produced above this melted mass, finding an outlet by the volcanoes, must necessarily carry away electricity, the same as the vapour formed above the sea, without there being any occasion to admit chemical subterranean effects producing electricity by means unknown to us.

248. If we now consider this emission of electricity in equatorial and tropical regions, where evaporation is very abundant there would naturally result clouds strongly electrified, and continual storms.

These clouds could not directly become raised to a great height; because they are carried away by the winds prevailing in these regions, and the electrical phenomena, appearing above the points where the storms have originated, continue to be produced in their passage, but become weaker as the latitude increases.

249. At the poles, on the contrary, where evaporation is far less rapid, the quantity of electricity tending to emanate from the earth is doubtless much less abundant, for the earth, being drier at the surface of the ground or the sea in these regions, becomes less easily charged; but that which is liberated rises directly into the higher strata of the atmosphere, and thus forms a kind of electrified curtain, tending to become diffused in space.

If no moist conducting mass happens to be interposed between this flow of electricity from the already elevated parts of the atmosphere towards still greater altitudes, the electricity becomes discharged in an invisible, or only slightly luminous, manner; for the transition from the less rarefied portions of the air to those more rarefied is not quick but gradual. Its passage does not then make itself known except by magnetic disturbances.

If, on the contrary, masses of clouds, in a condition of liquid globules or crystals of ice, float in the intermediate space, luminous effects are shewn, as in our experiments, and we see the polar aurora.[1]

250. This way of looking at the earth, as charged with positive electricity as well as the atmosphere itself, seems to render inexplicable, at first sight, the discharges which take place in ordinary storms between the clouds positively electrified and the equally positively electrified earth.

But this apparent difficulty is easily resolved if we consider that a given portion of the surface of the earth, although emitting positive electricity, is much less charged with it than the cloudy mass which passes above, after having gathered and stored in its course the positive electricity spread throughout the atmosphere, and also bringing a large part of that taken, during the formation of the cloud itself, above the sea in the warm regions.

The consequence is that this part of the ground, having only a comparitively weak positive tendency, becomes strongly negative by induction. It is the same as between the clouds themselves, which may be all positively electrified and yet be the seat of powerful discharges, because they are electrified in different degrees.[2]

---

[1] " The more recent observers place the seat of these phenomena, not at the limit of our atmosphere, but in the region where clouds and vesicular vapour are formed." (V. Cosmos, par A. de Humboldt ; traduction de MM. Faye et Galuski, 4e édit., t. I, p. 224).

[2] Messieurs Quetelet and Palmieri have allowed, as is known, that clouds which appear negative are only so by induction, and only at one of their extremities.

251. The arguments we have just set forth agree also, up to a certain point, with Ampère's hypothesis, admitting the existence of an electric current in a fixed direction, encircling the earth and producing its magnetic action.

We need only add, in order to explain the emission of electricity into the atmosphere, and, consequently, into space, that it must be a current of very high tension, impossible to be confined in a conductor of limited material like the wire of a solenoid, but radiating all round the mass of the earth on account of its high tension.[1]

252. Finally, we draw the conclusion from these premises, that the polar aurora results, in our opinion, from the diffusion of positive electricity, emanating from the polar regions themselves, in the higher strata of the atmosphere round the magnetic poles, either in invisible rays when there are no clouds interposed, or converted into heat and light by meeting with aqueous substances in either a liquid or solid state, which it vaporizes with a noise, and precipitates in the form of rain or snow upon the surface of the earth.[2]

---

[1] This paragraph and those which precede it from 244, respecting the origin of atmospheric electricity, are extracts from an article presented to the Académie des Sciences on the 13th March, 1876, from which an extract only was inserted in the Comptes rendus.

[2] Comptes rendus, t. LXXXII, p. 629, March 18th, 1876.

CHAPTER V.

Comparisons with Spiral Nebulæ.

———

· 253. If spectral analysis has permitted, in these later times, of the study of the chemical composition of celestial bodies, it is not chimerical at the present time to try to account for their physical constitution by the observation of the electrical phenomena and by the inferences to which these phenomena lead.

Herschell and Ampère had already thought that the incandescence of the sun might be attributed to electric currents. Several astronomers and modern scientists, among whom we may mention Messrs. Young, Morton, Respighi, Spœrer, Marco of Turin, have suggested similar ideas.

The phenomena we have observed with very high tension electric currents, such as the spiral motions, the luminous effects, the spherical or annular form taken by substances submitted to the action of the current, have led us also to think that electricity in a dynamic condition might play an important part, not only in meteorological phenomena, but also in those of celestial science.

The experiment described above (158), in which a cloud of metallic oxide, torn from an electrode by the current, takes a spiral motion in the body of a liquid under the influence of a magnet, seemed of a nature to explain, in particular, the remarkable form of spiral nebulæ.

It is, in fact, sufficient to take a glance at the figures shewing this experiment (figs. 65 and 66, page 142), in order to at once recognise the form of these nebulæ described by Lord Ross, some of which have the curve of their arms pointed in the opposite direction to the hands of a watch, as in fig. 65, such as the nebula of the virgin ("Chevelure de Bérénice"), &c.; others have their spiral turns directed the same way as the hands of a watch, such as those in fig. 66, and like the nebula "Chiens de chasse," etc.[1]

254. In view of so striking a similarity, may it not be reasonably supposed that the nucleus of these nebulæ may be formed by a veritable electrical furnace ; that their spiral form is probably caused by the presence of celestial bodies powerfully magnetised, and that the direction of the curve of the turns in the spiral must depend upon the nature of the magnetic pole turned towards the nebula[2]?

It would then be interesting to search, among the stars already known round these nebulæ, those which, by their position, could exercise this magnetic influence, or to explore the celestial firmament, on the axis of which the spirals appear to turn, within

---

(1) V. Le Ciel, par. Amédée Guillemin. 5e édit., p. 833 et suiv.

(2) These inferences may be objected to on the score that no conductor can be perceived in space leading an exterior electric current to the centre of the nebulæ. In answer to this objection we will call to mind that, among other experiments made with a far weaker source of electricity, we have noticed small luminous rings composed of incandescent particles entirely detached from the electrode; these rings, the centre of which is disturbed by a whirlpool, are set in motion in the space comprised between the electrode and the larger luminous ring formed around by the shock of the electric wave against the sides of the voltameter (157). They are in this case veritable electric fires, separated from the principal source which gave rise to them, and similar to, although infinitely smaller than, the nucleus of isolated stars, such as those which form annular nebulæ.

or beyond the plan according to which they become developed, in order to discover celestial bodies capable of causing their form or spiral motions.[1]

In the case where a star might be known to fulfil these conditions, we might still examine, along the line passing through the centre of the nebula and the star itself, if there were not a second spiral nebula connected with the other magnetic pole of this star, the curves of which, turning in the opposite direction to the magnetic currents of this pole, might, nevertheless, appear to the observer to be directed in the same direction as those of the first, and the combination of these three bodies would thus form a symetrical stella system.

Cosmic matter is spread with such profusion throughout space that this hypothesis partakes in no way of the impossible.[2]

As such research requires the use of the most powerful telescopes, we only beg to draw the attention of astronomers thereto with every reserve which inductions based upon simple analogies demand; but, if actual observation happened to justify them, it would assuredly be a decisive proof in favour of the electrical constitution of the celestial bodies.[3]

---

[1] Reference may also be made to electro-dynamic actions of this kind in order to explain the breaking up of celestial bodies as supposed by Mr. D. Vaughan of Cincinnati (July, 1878). This breaking up may also be facilitated, or begin, by the cooling of the stars being carried to its utmost limit, as supposed by M. Stanislas Meunier, in order to explain the origin of meteors. (Le Ciel géologique, p. 195, 1871).

[2] The nebula of the "Chiens de Chasse" itself affords another nebula which has been recognised by Chacornac as also shewing a spiral form.

[3] Comptes rendus, t. LXXXI, p. 749, October 26th, 1875.

CHAPTER VI.

Analogy with the Solar Spots.—Upon the physical formation of the Sun.

---

255. The electrical perforations made by high tension currents (160), figs. 68 to 70, appeared to us to present some remarkable similarity to the formation of the solar spots, such as those observed by Messrs. Nasmyth, Dawes, Lockyer, Chacornac, le P. Secchi, Tacchini, Langley, etc., and which have been compared with sprigs or bundles of thatch, and filaments, crooked, twisted, or entwined, etc.

The material in action in our experiments is very different, it is true. In this case it is simply moist organic matter, whereas, in the sun, it is a question, no doubt, of a fluid and incandescent mineral substance. But the action of electricity might take place, and shew itself in the same manner, in both cases: in the first, by sub-division of the dried up matter and jets of water vapour into innumerable threads; in the second case, by extremely finely divided streams of luminous matter, and by ejections of the vapour from mineral substances.

This curious appearance of spots on the sun, so difficult to account for by ordinary mechanical action, is easily explained by the intervention of electricity; one of the most characteristic properties of which is to form into points, or to cut up into threads, all matter opposing its passage in order to open up the many tracks which seem necessary for its rapid discharge.

256. We have thus been led to think that the spots upon the sun are cavities produced by essentially electrical eruptions; that, in consequence, the interior mass of the sun must be strongly charged with electricity; and that, according to the direction of the excavations, the thread-like slope of which turns towards the interior of the star, the electricity which escapes must be positive (160).

257. But these are not the only analogies which we can refer to in order to prove the presence of electricity in the sun. The phenomena observed in incandescent metallic globules, formed under the action of a powerful electric current, may also throw some light upon the physical formation of the sun, considered simply as a globe of incandescent matter.[1]

---

(1) Le Verrier wrote in 1860, after having observed a total eclipse of the sun in Spain:— "I have got at the physical formation of the sun. It seems to me that we must completely abandon the present theory. The thing may be much simplified."

We used to be told that the sun was composed of a dark central globe; that above this globe there was a deep atmosphere of dark clouds; higher still there was supposed to be phosphorus in the form of a gaseous envelope, luminous in itself, and the source of the light and heat of the sun. When this substance became torn, it was said that the dark centre of the sun could be perceived, hence the spots which frequently show themselves. To this complex formation it was necessary to add a third envelope, formed of a combination of roseate clouds.

Now, I fear that the greater part of these envelopes are the work of imagination, that the sun is simply a luminous body by reason of its high temperature, and covered by a continuous layer of the roseate matter, the existence of which we are at present aware. The star, thus formed of a central liquid or solid body, covered by an atmosphere, follows the law common to the constitution of celestial bodies.

Whatever the formation of the heart of the sun may be, solid or liquid, the surface and the interior of the star must be at least as much disturbed as the surface and interior of the earth, and neither waterspouts, electrical phenomena, nor volcanoes should be wanting, capable of producing the movements observed." (Bulletin de l'Association Scientifique de France, p. 97, 1869).

We have, in fact, seen that these melted globules (fig. 13 and 14, page 49), shew brilliant eruptions in consequence of the disturbance in their interior substance under both the calorific and chemical action of electricity, that the jets of gas and incandescent particles arising from these interruptions necessarily made their way out by the cavities or perforations produced in the interior of the globule itself; and that these perforations, allowing the comparitively colder and less luminous interior of the globule to be seen, formed dark spots upon its brilliant and undulating surface, and that, in consequence of the greater or less thinness of the melted envelope around the craters, some parts of the surface appeared more or less brilliant in the neighbourhood of the spots; that these globules, examined after cooling, shewed a wrinkled and pimpled surface; finally, that they were hollow and that their envelope was proportionately thinner as the metal enclosed more gas in combination.

258. We have then arrived at the following conclusions from these experiments by way of comparison :—

First, that the sun may be considered as a hollow electrified globe, full of gas and vapour, covered with an envelope of melted incandescent matter.[1]

Secondly, that the wrinkles or luculæ on its surface arise from the undulations of this liquified envelope.

Thirdly, that the spots are simply perforations in the fluid envelope, produced by quantities of gas and electrified vapour, emanating from the interior of the star, and giving to the sides of

---

(1) This conclusion agrees with the known low density of the sun.

the cavities, as above described, the formation which characterises the passage of positive electricity.

Fourthly, that the faculæ appear to be a brilliant phase in the evolution of the gaseous matters, when they approach the surface previous to their eruption.

Fifthly, that the protuberances are formed by the gases themselves escaping from the interior of the star at a higher temperature, and, consequently, more luminous than those forming the atmosphere upon its surface.

259. It may be objected that the metallic globules in question are produced between the two poles of an apparatus and traversed by an electric current, while the sun is isolated in space; but referring to our former experiments, such as that of the *spray* (143), one can understand the formation of electrified spheroids, totally detached from the source whence they arose, and carrying with them a considerable quantity of high tension electricity. Besides, if, in the experiment of the metallic globule, we allow the wire, to which the globule adheres, to melt, the current is broken; the globule remains suspended from one of the poles, and, during the brief moment that it remains incandescent, spots are still produced, and bubbles are liberated upon its surface (fig. 14, page 49).

If this phenomenon lasts an appreciable length of time with so small a mass it may be understood what duration it might possess when the gigantic globe of the sun is in question. The vibratory electric motion communicated to it must last like mechanical motion as long as its own physical and chemical effects.

Thus, we believe that the sun is electrified, but that it does not create the electricity which it possesses, any more than the heat and light which arise from it; it is a store received from the nebulous ring, of which it is but a brilliant particle, condemned to become extinct some day; this nebulous ring would be derived from another electrified wave and so on up to the primary cause, the creator of all power and motion. Taken from this point of view, the incandescence of the solar globe, prolonged through a series of ages, would be in itself but a spark of short duration in the infinitude of time and space.[1]

---

[1] Comptes rendus, t. LXXXII, p. 816, April 10th, 1876.

FIFTH PART.

Rheostatic Machine.

260. Description of the rheostatic machine.—After having described apparatus suitable for accumulating and transforming the work of the voltaic battery, so as to obtain, at will, temporary effects of quantity or tension very much higher than those of a given battery, and after having specially applied them to the observation of effects produced by electric currents of high tension, and having seen how much interest the study of these effects created by their analogy with natural phenomena, we have sought to more completely transform the energy of the voltaic battery, and to obtain an E.M.F. equal to that of electrical machines.

This problem seemed already solved, no doubt, by induction apparatus; but the method we used, although less simple from a practical point of view, appeared to us more direct in theory, and capable of converting, with less loss in transformation, a given quantity of electric energy ready to provide a dynamic current, corresponding in quantity with electrical effects of the static kind.

We have already had occasion to frequently prove that our secondary batteries of 600 to 800 elements permitted of the rapid charge of a condenser having a sufficiently thin insulating plate of glass, mica, gutta-percha, paraffin, &c.[1]

Fig. 74.

In order to obtain continuous static effects of the greatest intensity, we joined a certain number of condensers, formed, by preference, of mica coated with tin foil, arranged like the secondary batteries themselves, so as to be successively charged in parallel and discharged in series; and we called the apparatus thus invented, by the name of rheostatic machine, (fig. 74).[2]

261. All the plates of this apparatus must be insulated with great care. The commutator is formed of a long vulcanite cylinder furnished with longitudinal metal strips intended to connect the condensers in parallel, and pierced transversely by

---

(1) We know that Volta, Ritter, Cruikshank, etc., were able to charge condensers by means of the battery, and that these results gave occasion for subsequent research, by calculation or experiments, on the part of many scientists.

(2) Comptes rendus, t, LXXXV, p. 794, October 29th, 1877.

copper wires bent at each end, or with metal hooks slightly rounded, for the purpose of uniting the condensers in series. Wires in the shape of springs rr' are connected with the two armatures of each condenser and fixed upon a bar of ebonite on either side of the cylinder which may be rotated rapidly by means of the wheel R and a pinion geared with it.

When the cylinder is turned so as to present its longitudinal metal strips in contact with the springs, the even row of all the condenser armatures are united on one side and the odd row are united on the other side, so as to form a single condenser of large surface, charged by connecting the terminals P and P' to the poles of the battery.

When the cylinder is turned the other way, as shown in the figure, so as to present the transverse wires to the springs, all the charged condensers become united in series, or tension. The armature L of the condenser on the extreme left communicates with the last spring on the far side of the cylinder, and terminates at the arm E of the exciter. The armature L' of the last condenser on the right communicates with the last spring but one on the next side; this spring comes in contact with the last metal pin which pierces the cylinder from one side to the other, and the last spring on the far side of the cylinder communicates with the other arm E of the exciter.

Whilst the condensers are thus united in series, the battery charging the apparatus is entirely thrown out of the circuit; the last spring r', seen in the figure, communicating with the battery through the terminal P', and the other end spring communicating with the terminal P, do not touch any metallic portion of the commutator cylinder.

262. Effects produced by the rheostatic machine.[1]—
To study the effects produced, we have used several machines, composed of a varying number of condensers having a double armature of about three square decimetres in surface.

By first employing a machine with six condensers, charged by the secondary battery of 800 cells, we obtained, at a speed of fifteen revolutions per second, a series of brilliant sparks, from thirteen to fourteen millimetres in length, following one another at the rate of thirty per second, and forming a continuous stream of fire, accompanied with the same noise as that of the sparks from an induction coil in connection with a Leyden jar.

With machines composed of thirty, forty and fifty condensers, we have obtained sparks of four and five centimetres.

263. The length of the sparks increases nearly in direct ratio with the number of condensers; but it is not possible to establish it definitely on account of the inequality in the thickness of the insulating plates and the variable effects resulting from it. Thus, with a series of thirty selected condensers, with very thin insulating plates, the sparks are four centimetres; with fifty condensers of various thickness, they hardly exceed five centimetres.

264. On the other hand, when the insulating plate of the condensers is too thin, it is pierced by the current, and we then obtain, if the current continues to act, the *wandering spark* described above (fig. 48), p. 125).

265. The diameter of the commutator cylinder must be proportioned with regard to the length of the sparks expected to

---

[1] Comptes rendus, p. 761, March 25th, 1878.

be obtained. One quarter of its circumference, or the interval comprised between the metal strips of the commutator and the line of transverse pins, must be greater than the length of the spark which can be created; otherwise the latter would strike across the commutator instead of appearing between the arms of the exciter.

266. **Crooked and "brush" form of the discharge.**—The spark created by the rheostatic machine when a constant and sufficient speed of rotation, and E.M.F. of the primary current, is used, presents a special and very regular form that is not seen with the same degree of clearness in that from electrical machines or induction coils.

Fig. 75.

This form, when the angle comprised by the arms of the exciter is very obtuse, consists in a stream of fire starting in a line with the positive arm, and rising noticeably above the negative point, to which it returns in the form of a crook, making numerous deviations in descending upon this point (fig. 75).

When the distance between the points is increased by one or two millimetres, the discharge given by the machine takes the form of a brush, following the same course as the former one. A luminous spray springs forth from the positive pole for about three quarters of the distance between the poles and turns back towards the short brush formed round the negative point (fig. 75).

267. The difference in the form of these sparks and "brushes," compared with those from induction coils, is accounted for by the rheostatic machine not giving, like these latter apparatus, an alternating flow of electricity, but always in the same direction, which allows, besides, of its easy measurement with Thomson's long scale electrometer for tension, and of its comparison with that of electrical machines.

268. There is also occasion to remark, as we have above foreseen (260), that the loss of energy resulting from the transformation of dynamic into static electricity is, in this case, much less than in induction apparatus; because, the voltaic circuit not being for a moment short circuited, there is not that conversion of a part of the current into heat. The current simply spreads over the polarising surfaces presented by all the condensers in proportion as they are discharged.

269. With machines of thirty to fifty condensers, giving sparks from four to five centimetres long, and the speed of rotation not being so great, the sparks are less continuous, and their form is not so regular; their deviations arise or descend irregularly above or below the straight line between the two points of the exciter (fig. 76).

The positive "brush" then presents a calyx terminated by a luminous ovoid sheaf, more or less branching out, just like that from electrical machines (fig. 76).

270. Light produced in vacuum is brighter, in this case, than from electrical machines, in consequence of the greater quantity of electricity used, and, when the rotation is quick enough, it is as bright and continuous as from induction coils.

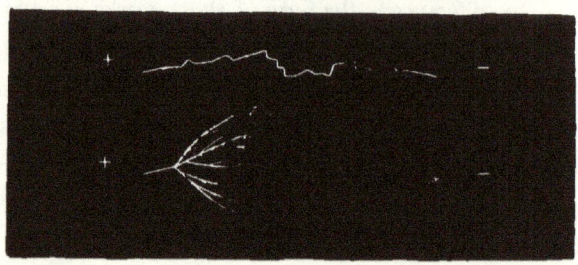

Fig. 76.

Geissler tubes of the highest resistance, and M. Ed. Becquerel's tubes of phosphorescent substances, are illuminated in a brilliant manner.

The absence of all stratification, and of the sheath of blue light round the negative pole presented by Ruhmkorff coils, is noticed. The light is of a purple colour throughout the tubes, and completely fills them in the same way as with an induction coil and Leyden jar.

This effect must arise from an excess of tension; for, if that of the battery used to charge the rheostatic machine be much reduced, the blue sheath and stratifications appear.

271. **Light in a Vacuum.**—The secondary battery of eight hundred cells used to charge a rheostatic machine, although not specially well insulated, is able to illumine Geissler tubes directly, by producing in them stratifications such as those observed by Messrs. Abria, Grove, Gassiot, Warren de la Rue and

H. W. Müller. A long column of water being introduced in the circuit, we can, with a single discharge of the battery, make a Giessler tube luminous for more than three hours and a half, on account of the very small quantity of electricity expended in the passage of the current through the rarefied air.

But when the current from the battery is cut off whilst the tube is being illuminated, it often happens that it is not reproduced by reclosing the circuit. A slight diminution in the tension of the battery is sufficient to prevent the phenomenon from taking place; because the E.M.F. of eight hundred secondary cells is about the minimum tension that may be used to make electricity flow through narrow tubes of rarefied air.

If we then introduce the rheostatic machine into the circuit, so that the terminals of the vacuum tube communicate at the same time with the poles of the machine and of the secondary battery, we notice, in turning the machine for an instant only, the curious result of the tube being immediately illuminated without stratification, and of the battery continuing to illuminate it alone with stratification. The tube was "primed" by the E.M.F. of the rheostatic machine being greater than that of the battery.

272. In general the rheostatic machine will give all the other effects created by electrical machines and induction coils, and these effects do not seem to be disturbed by variations in the hygrometric condition of the air.

Production of the continuous spark or "brush" is accompanied by a strong smell of ozone. Each of the poles will give sparks when brought near to objects connected with the ground. The electric "whirl" or "vane," or the effect of blowing produced by the points of the exciter, are good examples.

273. **On the possibility of obtaining some effects with a current of less tension.**—The apparatus in question would offer but a theoretical interest, if it were always necessary to employ a battery of eight hundred cells in order to show its effects. We have consequently set to work to produce them with a far less source of electricity and we attained the desired result by increasing the number of condensers and decreasing as much as possible the thickness of the insulating plates.

With a machine of fifty condensers with very thin mica plates, held by frames of vulcanite, continuous sparks of six millimetres were obtained by only using one hundred secondary cells and we could even illuminate a vacuum tube by charging the machine with a secondary battery of thirty to forty cells.[1] It is with this relatively weak source that the blue sheath and striæ are visible round the negative pole.

274. **Complete transformation of a certain quantity of dynamic into static electricity.**—It was interesting to attempt to give an example of complete transformation, by means of the rheostatic machine, of a certain quantity of dynamic electricity stored by a secondary battery, and to know, approximately, the time required to completely exhaust the charge in static effects. Among the various experiments we have made may be mentioned the following:

---

[1] In place of a secondary battery, a Bunsen battery of from fifty to sixty elements, such as is set up for the electric light, or a Gramme machine wound for the highest possible E.M.F. would do equally well for charging the rheostatic machine.

We have even obtained, with a very little secondary battery of six hundred elements (formed of small forked plates of lead, of a few millimetres only in width) in connection with the rheostatic machine, sparks almost as long as with six hundred of our ordinary elements.

These six hundred little cells were immersed in glass tubes of one centimetre in diameter, fixed upon a board only sixty millimetres long by forty millimetres wide. All these cells were charged in series, by our large battery of eight hundred cells.

A secondary battery of forty cells, without any residual charge, but quite ready to store the smallest amount of chemical work of the primary battery, was charged for fifteen seconds only by two Bunsen elements and then coupled up to the rheostatic machine. It was necessary to turn the apparatus for more than a quarter of an hour to exhaust that charge in the illumination of a Geissler tube (v. § 3000).

It may be inferred from this that, with the quantity of electricity absorbed by a secondary battery in ten minutes (which is about the best time to accumulate without noticeable loss the work of the primary battery), we could maintain a vacuum tube luminous for more than ten hours.

275. **Length of the Sparks.**—We have seen above (263), that the length of the sparks produced by the rheostatic machine was clearly in proportion to the number of condensers. But the inequality in the thickness of the mica plates did not permit of establishing this in a positive manner.

By employing, since that time, condensers with mica plates of as uniform a thickness as possible, charged under the same conditions by the secondary battery of eight hundred cells, and by only making a half turn of the commutator of the machine, so as to produce but isolated sparks, instead of a continuous stream of fire, we have obtained some figures relatively higher; and, by charging more regularly than the preceding, with a machine of ten condensers, we have made sparks of a centimetre and a half; with a machine of thirty condensers, sparks of four and a half centimetres and with a machine of eighty condensers, sparks of twelve centimetres in length.[1]

---

(1) If we calculate the deviations described above, these sparks have a greater length; but we only in these cases measured the distance between the arms of the exciter.

The length of the sparks produced by the rheostatic machine may then be considered as proportionate to the number of condensers.

276. This length increases more rapidly with the E.M.F. of the current which acts upon the machine, and seems to vary in proportion to the square of the number of elements, the same as the direct spark from a battery of high tension, according to the law given by Messrs. Warren de la Rue, and Hugo W. Müller; but the results we have obtained were not always sufficiently consistent to enable us to affirm, with certainty, that the spark from the rheostatic machine exactly follows the same law.

277. **Large rheostatic machine.**—Figure 77 represents the rheostatic machine of eighty condensers that we used in these experiments. The vulcanite cylinder of the commutator is one metre long by fifteen centimetres in diameter.[1]

The arrangement is otherwise nearly the same as that of the first machine we described (261), but the end springs are at a sufficiently great distance from those which are next to them to prevent sparks striking across from the tension poles of the rheostatic machine to those of the secondary battery. The condensers are made with mica plates eighteen centimetres long by fourteen wide coated on each side with tin-foil. Fine copper wires covered with gutta-percha are fastened to the end of each armature. The edges of the condensers are also fixed in frames, or only plain ebonite plates, in order to give them more

---

[1] According to the rule we have given (265), it appears impossible to obtain, with this cylinder, sparks of a greater length than a quarter of the circumference (118 m.m.); but as the points of the exciter, placed opposite one another, offer an easier means of discharge than the space between the strips and pins on the commutator, the spark strikes across these points even when the distance is a little greater.

rigidity and maintain them more easily in a vertical position close to each other without contact.

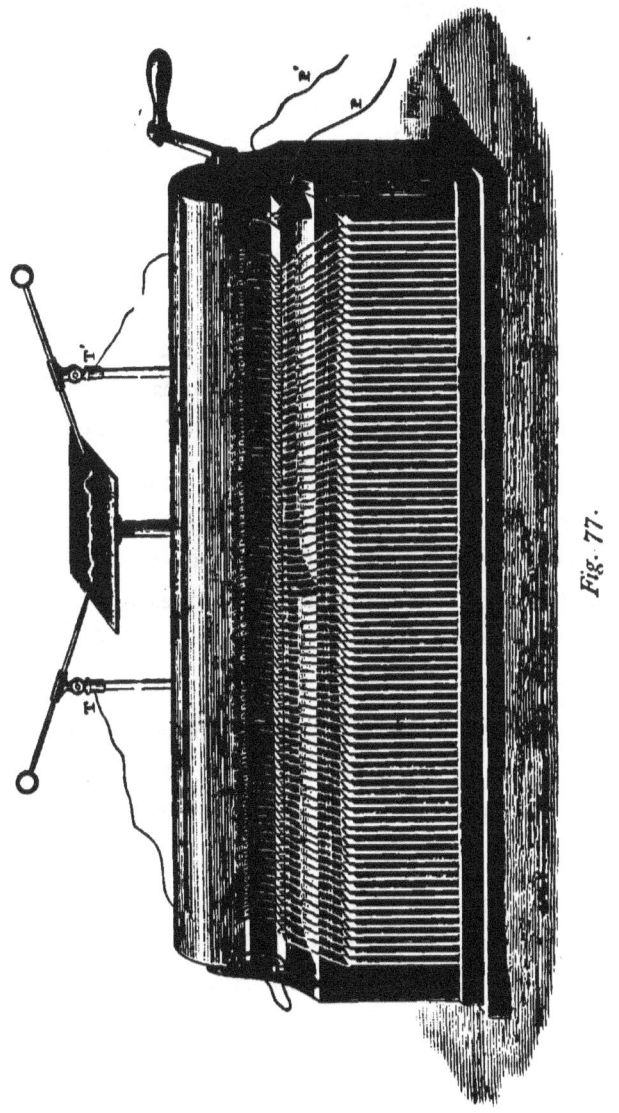

Fig. 77.

278. Upon rotating the commutator, sparks appear upon all the points where the metal strips meet the springs forming the terminals of the condensers for charging them in parallel, and give the cylinder the appearance of a sparkling tube.

Another time, sparks appear when all the condensers are united in series, and a discharge is produced between the arms of the exciter.

If a column of distilled water be placed in the circuit of the secondary battery, the water seems to be decomposed in a continuous manner whilst the machine is in motion. In reality, this decomposition only takes place at the moment when the sparks of *charging* are produced; for, during the discharge, the water tube, like the secondary battery, is thrown out of the circuit.

The limited quantity of dynamic electricity stored in the secondary battery is thus expended little by little during the charge only of the condensers; but this expenditure is very slow, and each charge of the condensers, and each discharge also, corresponds with a very small quantity of electro-chemical action consumed in the battery (300).

279. **Sparks produced under different conditions.**— We have seen (275) that sparks produced through the air by the rheostatic machine of 80 condensers, attained a length of 12 centimetres.

If flower of sulphur be spread between the two points of the exciter, supported upon a plate of insulating material, the space is rendered decidedly less in resistance and sparks may be thus obtained 15 centimetres in length (fig. 77). If we cause them to strike across a conducting powder, such as metal filings, they will increase to 70 centimetres.

280. When these sparks traverse flower of sulphur they make,

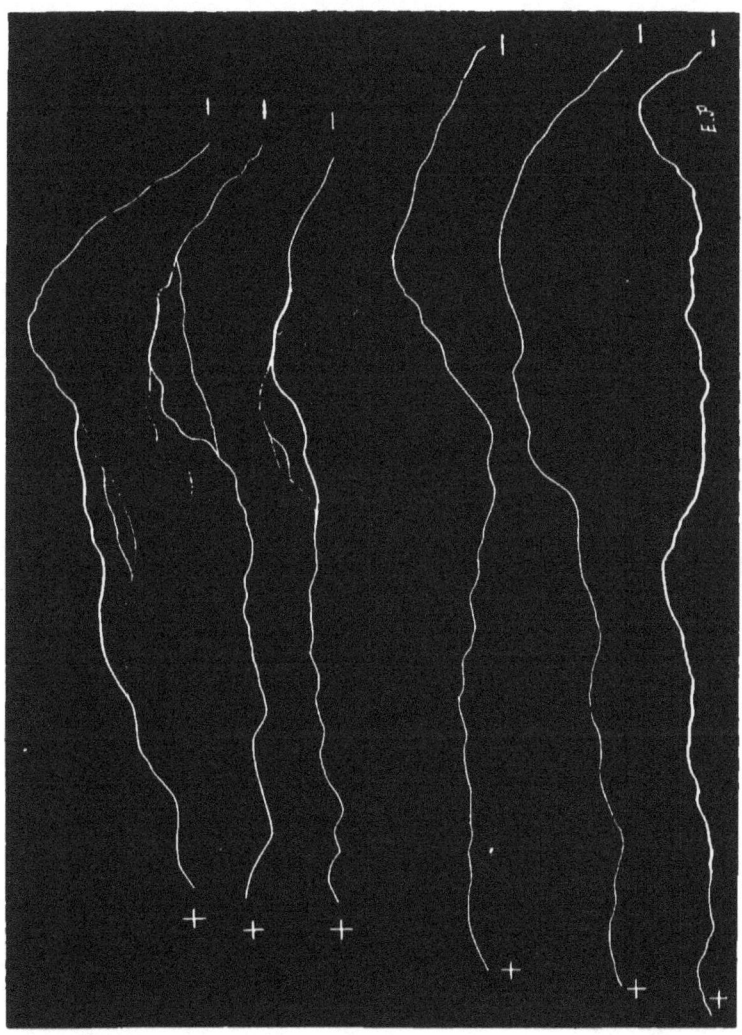

Fig. 78.

in their passage, a sinuous furrow from 2 to 3 m.m. wide, and, if the insulating surface, upon which the flower of sulphur is spread, consists of a mixture of resin with about one tenth parafine, they leave in the middle of the furrow a very clear bluish line visibly traced, with a leadish appearance, which permits of the preservation of the exact outline.[1] This trace is easily effaced, however, by rubbing, but by carefully following it and scratching it by means of a pointed instrument it is made ineffaceable and can then be easily copied.

It is thus that we obtained fig. 78, which represents sparks of various length the actual size.

281. **Form of the sparks.**—We notice that these sparks, when they have not attained their maximum length, often show closed branches which might escape notice when the luminous stream only is seen.

Their deviations are always rounded, and we never see that zig-zag form with sharp angles in which the electric sparks are often represented. The sinuous form predominates; sometimes the spark is reduced to two undulations, composing a kind of S, frequently found in lightning which strikes the ground (188).

282. The crooked form, in particular, which we have already described in the smaller sparks from a rheostatic machine of ten condensers, is found near to the negative pole (266). The shape in which the hook is produced constantly varies. Thus, we have obtained series of sparks in which the crook is in the opposite direction to that in fig. 78 but always near to the negative pole (fig. 79, 286).

---

[1] When the insulating surface is of pure resin or vulcanite the trace is not nearly so easily seen.

283. The formation of this crook appears to us capable of being explained by the meeting of the two ponderable matters, in motion from opposite directions, projected from the points of the exciter, and by the angle that nearly always results from this meeting; for the electric jets are hardly ever produced in a line with each other; they start from various points at the ends of the exciter however fine they may be. Each of these ends, even when pointed, offers in reality a relatively large surface in proportion to the extreme fineness of the jet of electric matter which escapes from it, and the point from which this jet strikes depends upon the most varied circumstances, either in physical condition or the chemical state of the surface, as we have had occasion to notice in the more abundant stream of luminous matter produced by a high tension current between a platinum electrode and the surface of a liquid.

As regards the formation of the crook close to the negative pole, this is accounted for by admitting that the electric movement starting from the positive pole must be the more rapid of the two, and that it traverses the greater part of the distance to the other pole whence there takes place an opposite movement; consequently, the angle or rounded crook resulting from the meeting must be naturally produced in the vicinity of the negative pole.

284. **Arborisations.**—These discharges also present arborescence, which appears on dispersing the excess of sulphur by giving the insulating plate a few light taps.

Figure 79 represents, in natural size, the arborisation formed along the course of a discharge fifteen centimetres in length, produced by the rheostatic machine.

285. These effects account for the vegetable appearance that

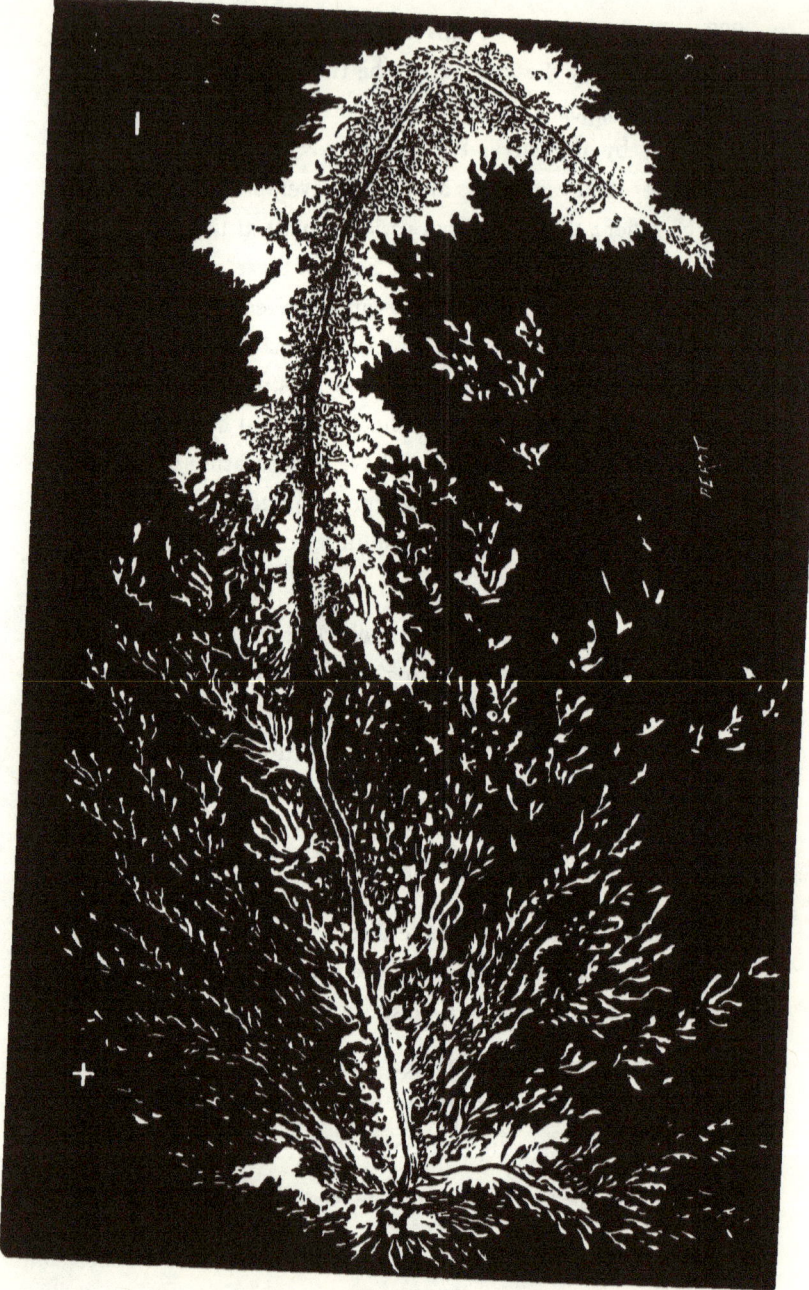

Fig. 79.

has been sometimes noticed imprinted upon the bodies of persons struck by lightning, which are only the result of ramifications in the track of the lightning itself.[1]

This case is easily explained by its analogy with what transpires in the preceding experiment. At the moment when the discharge is produced, we see the flowers of sulphur thrown into the air, especially round the two poles. In the same way, in the case of the lightning stroke, the dust or any matter placed in track of the discharge must be dispersed and we can imagine that this matter raised to a very high temperature, might produce upon the human body an effect of instantaneous cauterisation in an arborescent form.

286. Fig. 79 represents the track produced in flowers of sulphur by a spark from the rheostatic machine before giving the insulated plate, on which the powder is spread, the shaking which makes the arborisations appear.

It may be remarked that the breadth of the track is greater in the direction of the positive and narrows as it approaches the negative pole.

Round the positive pole may be perceived some jets corresponding with the branches or rays along which the flowers of sulphur has been thrown up and scattered in greater quantity.

---

(1) An instance of this was recently given in the "*Lancet:*'——"A shepherd in Leicestershire was watching his flocks in the fields when a storm broke out and, naturally, like most people insist on doing, he sought refuge under a tree. A short time after, he felt a shock on his left shoulder and, suddenly losing the use of his legs, fell to the ground. When he was carried home he was still in complete possession of his senses but he complained of pains in the back and the legs. The examination to which he was subjected by the doctor who was called in, brought to light a rather odd effect of the lightning. From the left shoulder, for the entire length of his back, there appeared, wonderfully produced, irregularities upon the skin, and, in a brilliant scarlet tint, a tree-like stem with numerous branches delicately traced as with the point of a needle. The trunk was about three-quarters of an inch wide and the general appearance was that of the lower part of a fern with six or eight branches. The whole was very well reproduced as if printed upon the patient's back." (Les Mondes, September 12th, 1878).

Near to the negative pole, on the contrary, circular tracings are seen, corresponding with the outline of the arborescent bouquets formed around this pole, which are, as has been shown in fig. 80, of quite a different kind from those near the positive pole.

287. Lichtenberg figures, produced by sparks from the rheostatic machine.—These sparks, produced on a surface of pure resin, by means of blowing the powdered sulphur and red

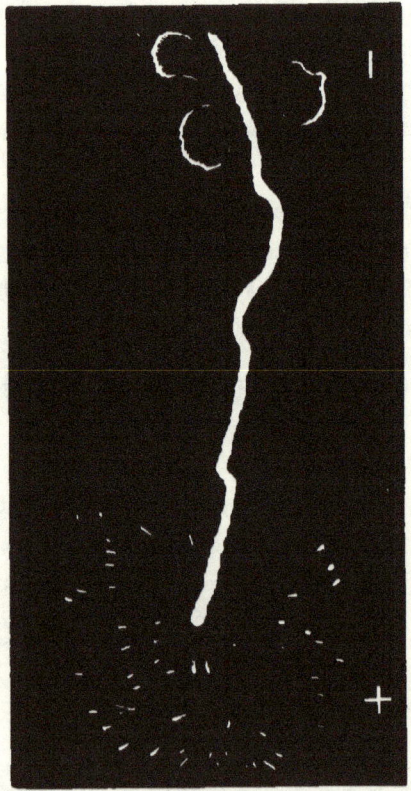

Fig. 80.

lead, produce good figures like those of Lichtenberg, of a different kind from the arborizations above mentioned and which, fixed upon paper moistened with varnish, constitute valuable examples for the study of the electric discharge (figs. *81 and 82). The difference between the effects produced by the sprays and that produced by sparks is here specially marked.

When the distance between the points of the exciter is too great for the spark to strike across and only a spray appears, the electric motion of ponderable matter leaving the negative pole, shown by the powdered minium adhering to the resin, does not reach the positive pole. This latter pole shows no trace of red dust in the midst of the crown of sulphur with diverging rays which surrounds it (fig. 81).

But, if the spark has struck across, this crown is open and the interior becomes filled with dust of red lead, showing that the electric motion leaving the negative pole has reached the point from which the positive electricity starts, fig. 82.

In the case of the spark, the distribution of negative electricity presents a curious crabiform appearance (fig. 82)[1]; in that of the spray, the electric motion around this same negative pole offers an aspect, no less strange, of a polypus directing its tentacles towards the positive pole without being able to reach it (fig. 81).

288. Moreover, when the spark flashes, one may sometimes know by the traces of sulphur around the negative pole, that the emission of positive electricity has extended even so far as to reach that pole. There is then a mixture of the two electricities at each pole (fig. 83).

---

*Fig. 81 reduced to three quarters natural size.

[1] This form is not exceptional; it has occurred in a great number of sparks obtained in the same manner,

Fig. 81.

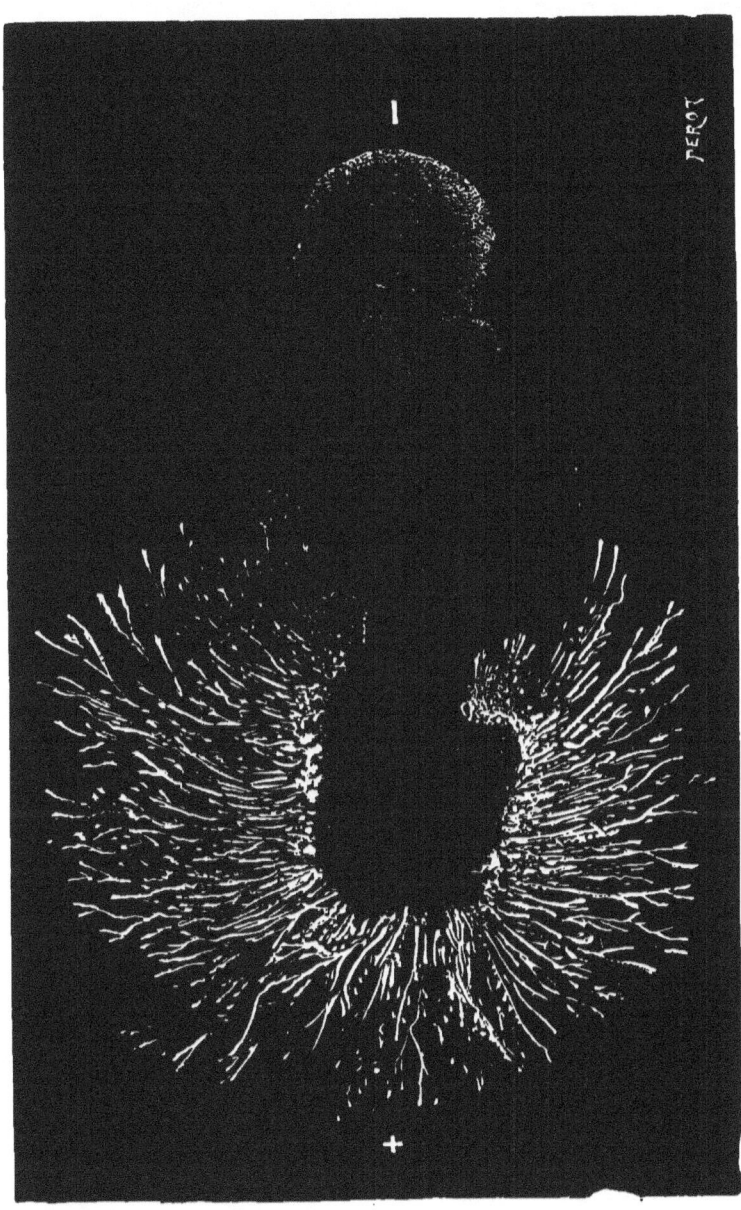

Fig. 82.

289. This observation explains how, in the circuit of currents of very high tension which closely resemble a continued series of discharges of static electricity, a complete decomposition of the water at each pole may be obtained and, consequently, a mixture of oxygen and hydrogen (132).

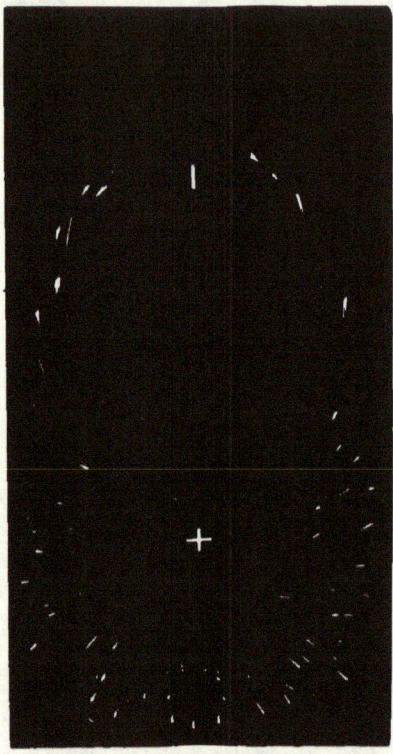

Fig. 83.

290. It may be also seen that the movement leaving the positive pole surrounds the negative electric movement as with a bunch of squibs of a curved trajectory (figs. 81, 82 & 83).

There is often seen, at the same time, an interior flow of positive electricity round the line of the spark, besides the positive current surrounding the exterior, and, between the two, the electric negative current which seems as though it were drawn in by the positive pole (fig. 84). Negative electricity, or the ponderable matter which it carries with it,[1] moves in an annular space formed by the electrified matter coming from the positive pole.[2]

291. This latter fact explains the effects of suction or ascension of water that we have obtained with electric currents of high tension (148).

Does it not also explain, as we have already pointed out, the ascension of water in the cloudy substance of water spouts (228)?

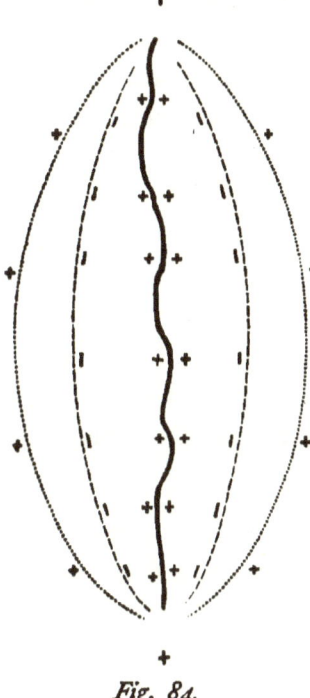

Fig. 84.

292. **Condensation of the Sparks.**—Sparks from the rheostatic machine make a considerable noise; for they result from the discharge of a very great number of condensers. But their intensity may be further increased by charging Leyden jars or

---

(1) When we say ponderable matter, we do not mean powdered sulphur and mimium, but that matter, invisible when cold, drawn from the electrodes by disruptive discharge, the passage of which on the resin is revealed by the subsequent insufflation of the mixture of sulphur and red lead.

(2) All these effects may without doubt be observed by means of discharges from Leyden batteries; but we here describe them as obtained with the rheostatic machine, not only to show their identity with those of static electricity, but also because this machine allows us to reproduce them on a large scale and with greater clearness than by any other means.

batteries. It is only necessary to leave an interval of one or two centimetres between one of the poles of the machine and one of the armatures, and letting each charging spark flash across the space, as with induction coils, in order to prevent any partial discharge from the jar or battery when the condensers are in parallel during the rotation of the machine.

Under these conditions jars or batteries may be charged, just the same as with a powerful electrical machine, and retain their charge very well.

Five or six sparks suffice for highly charging a large jar. A battery composed of four bottles is charged in a few seconds; for, even by a very quick rotation of the machine a continued series of powerful sparks is obtained.

293. In using the Lane electrometer, so that the jar discharges spontaneously and in proportion as the machine rotates, frequent sparks of 5 centimetres in length strike between the branches of the exciter with a loud noise.

294. **Coloured sparks.**—Sometimes, under certain conditions, the sparks are seen to be of a yellow colour much brighter than the ordinary sparks and analogous to those which have been obtained by M. Teploff with an electrical machine.[1]

These are the circumstances under which we have remarked them. The circuit of the rheostatic machine, when in series, is closed by a Leyden jar of very small surface, formed by a long tube of thick glass that a single spark from the machine is more than sufficient to saturate, and the distance between the armatures

---

(1) Journal de Physique, t. VIII, p. 131, 1879.

of which (equal to $0^m 20$) is such that the discharge cannot take place.[1]   Besides, the exterior armature is formed of two metal rings about a centimetre apart.

Electric brushes then appear between the first ring of the outer armature and the unattached rod outside which communicates with the inner armature.   But, at the same time, in the space between the inner armature and the second ring of the outer armature, are seen the yellow sparks of which we speak. These make little noise; they seem rather to flash between the interior and exterior of the glass.

295.   This kind of spark seems to us to be owing to a partial or imperfect discharge across the glass, and to the electro chemical action which results from it.

The discharge is not strong enough to pierce the insulating substance,[2] and, on the other hand, the tension is such (the armatures being charged to saturation) that the electric decomposition must be accomplished more or less completely in some way or other.   The result under these conditions is an electrical effect of *quantity* condensed on one point, an effect in this case rather calorific than mechanical, consequently, a chemical decomposition of an extremely small part of the interposed substance, and, as this substance is of glass, an incandescence of the sodium which gives the spark its characteristic yellow colour.

296.   The red colour observed by M. Teploff with moist conductors such as wet cords ought to be explained, in our opinion,

---

(1)   If this distance is only from $0^m 14$ to $0^m 15$ the spark flashes between the armatures.

(2)   The conditions are so arranged that this effect cannot take place, for two sparks are sometimes sufficient to pierce a jar of small surface formed by a tube of thick glass.   It would be possible nevertheless that a very fine crack or an invisible hole might be produced.

in the same way, by the incandescence of a very small quantity of released hydrogen.

In short, glass or other substances would not act here only as non-conductors but as electrolytes on the surface.

297. We have obtained, in addition, sparks of a red colour, by charging, by means of the rheostatic machine, condensers with thin insulated plates of vulcanite. One or two sparks suffice to pierce them; and the sparks resulting from a partial charge, which is subsequently obtained, are seen in the form of a little cylindrical red flame emerging from the two sides of the hole in the condenser and passing about half a centimetre beyond it.

The red colour is, in this case, again owing, in our opinion, to the hydrogen from the vulcanite, partially decomposed by the passage of the discharge under these particular conditions. The extreme tenuity of the hole formed allows of a certain accumulation of electricity on the armatures and also its egress in a large enough quantity to produce electro-chemical action.

298. **Halo of red light in a vacuum.**—Sparks from the rheostatic machine easily illumine Geissler tubes, as already mentioned (270), without producing stratifications, unless the current which works the machine be reduced to a relatively low tension.

The light surrounds the two electrodes and more completely fills the tube than the light supplied directly by the 800 secondary couples. There is no marked difference in its appearance at the two poles.

But, if air, rarefied by the condensed spark of which we have just spoken, be made to pass through the tube (293) by placing the latter between the arms of the exciter, in the circuit of which is placed a Leyden jar, it may be observed that the light produced at the positive pole is encircled by halo or fringes of bright red.

299. We think this colouring of light in a vacuum may be explained by the incandescence of hydrogen proceeding from the small quantity of water vapour of which the glass tube always contains some traces.

When the spark from the machine (when is series) passes through the tube, the decomposition does not happen in a perceptible manner because the *quantity* of electricity is wanting. But, as soon as there is condensation, electro-chemical action is produced and the colouring peculiar to incandescent hydrogen appears.

300. **Means of valuing very small quantities of matter or very short intervals of time by means of the rheostatic machine.**—We have already described (274) an experiment intended to show that, by means of the rheostatic machine, a small quantity of dynamic electricity can be completely exhausted in the form of static effects. A secondary battery was charged for only fifteen seconds; then it was made to work on the machine and this feeble charge, when transformed, could illumine a vacuum tube continuously for more than a quarter of an hour.

According to the number of turns made by the machine and the number of sparks obtained at each turn of the commutator, it is found that the action of the primary cells on the secondary battery during fifteen seconds, corresponds with the production of about 10,800 sparks in a vacuum.

It follows that the action of the battery during one second is represented in this experiment by 720 sparks from the rheostatic machine, or, in other words, the production of a spark corresponds to a duration in the action of the primary battery of $\frac{1}{10}$th of a second.

On the other hand, by introducing a voltameter into the circuit of the primary battery, we have discovered that the charge taken by the battery and yielded by the rheostatic machine in the form of those 10,800 sparks, corresponds with a consumption of 18 milligrammes of zinc in the primary battery.

It follows that the solution, or deposit, of one milligram of metal may be thus proved by the production of about 600 sparks from the rheostatic machine, or, in other words, that the production of one spark corresponds with the consumption of about $\frac{1}{600}$th of a milligram of metal in the primary battery.

Then, by taking the electric sparks as units, there would be a means of measuring either very small quantities of metallic matter dissolved, or deposited, by electro-chemical processes in very short intervals of time.

It is, besides, easy to determine the number of sparks produced in air or in vacuum by the rheostatic machine; because we know the exact number of sparks made by each turn of the machine. The machine may be also made to turn as slowly as one wishes and to expend in any length of time, in a static form, the effect produced in an extremely short time by dynamic electricity.

301. We may likewise infer from these experiments that the feeble static tension effects shown directly by the poles of a battery composed of a great number of elements must not be considered, as was at first thought,[1] as independent of the electro-chemical

---

[1] See Gassiot, Philosophical Transactions, 1844.

action produced in the battery; on the contrary, they correspond with real electro-chemical expenditure, doubtless very small, still not absolutely nil, even when the circuit seems open and it is question of but a single spark given with the electroscopic condenser, or of an act of simple attraction or repulsion.[1]

302. Effects of quantity from the rheostatic machine. The rheostatic machine can also give static effects of *quantity* which materially differ from those of *intensity*, by combining all the condensers in parallel and fitting it with another small special commutator intended to collect the discharges without confusing them with the effects of the secondary battery.

This commutator is formed by a little cylinder in vulcanite on which are four copper bands as at $m\,n$ and $o\,p$, placed in pairs opposite each other, against which the six springs B C E B'C'E', rub, also in pairs, opposed to each other (fig. 85, 86).

The pair of springs B B' communicate with the secondary battery; the pair C C' with the two charging poles of the rheostatic machine above described (260, fig. 74), previously turned into a position so that the condensers shall be combined in parallel. The pair of springs E E' communicate with an exciter or any other apparatus through which the discharges are required to pass.

---

(1) It is difficult to conceive at first sight the production of electro-chemical effect in a battery the circuit of which is not completely closed. But it must be considered that when the two poles of a battery are put in communication with the armatures of an electroscopic condenser, it exercises across the insulating substance a peculiar action equivalent to an imperfect passage, which necessarily involves a chemical expenditure in the battery. We have seen, in fact, that if the battery is at a high tension, the charging sparks are perceived at the moment when the battery is put in connection with a system of condensers and it can be even proved that there is a decomposition of the water in a voltameter placed in the same circuit as the condensers (278). These effects show that there is a passage (to a certain degree) of the current from the battery; accordingly, it must produce in the interior a corresponding chemical action. It is the same in a lesser degree if one of the poles of the battery is put in connection with an electroscopic condenser, the other pole being in connection with the ground, or even if a substance touching the ground is only brought near to one of the poles, for there is always an interval between the poles of the battery filled with the surrounding air which plays the part of the dielectric medium in a condenser.

In the position of the commutator represented by fig. 85 the two poles of the battery B B' are connected by tongues, as at *m n*, with the two poles of the rheostatic machine C C', and the condensers then become charged simultaneously in parallel.

In the position of the commutator represented by fig. 86 the poles of the battery are outside any circuit and the poles of the rheostatic machine, charged in parallel, communicate by the two opposite tongues, as at *o p*, with the springs E E' in the circuit of discharge.

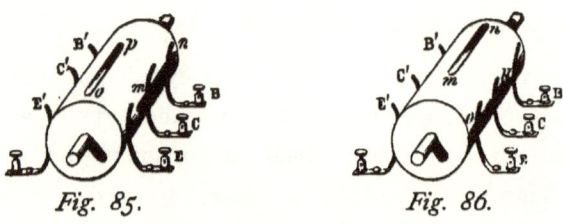

Fig. 85.   *Fig. 86.*

This commutator may be rapidly rotated in order to produce a nearly continuous series of static discharges in parallel for *quantity*.

Thus, whilst in the rheostatic machine before described all the condensers are charged in parallel and discharged in series, in this case the condensers are charged and also discharged in parallel, immediately after their charge, by the secondary battery without the current from that battery interfering in any way with the circuit of discharge from the condensers.[1]

---

[1] Among the experiments or arrangements of apparatus with which the rheostatic machine when arranged for quantity may be analogous we mention those of M. Werner Siemens who had obtained permanent galvanometrical deviations with condensers composed of different insulating substances successively charged with the use of a vibrating balance by some Daniell elements and had determined in this manner the inductive power of these substances. (See G. Wiedemann, Galvanismus, 2nd edition, vol. I, p. 199. — E. Mascart, Traité d'électricité statique, vol. II, p. 400).

303. The commutator which we have just described, instead of being arranged with a special driving gear near to the rheostatic machine, may be fitted to the machine itself, as at $a'b'$ (fig. 87) so that it may be put in movement by the rotation of the machine independently of the commutator for quantity and tension, $a\ b$ being kept at rest in a position which unites in parallel all the condensers by pressing the button K.

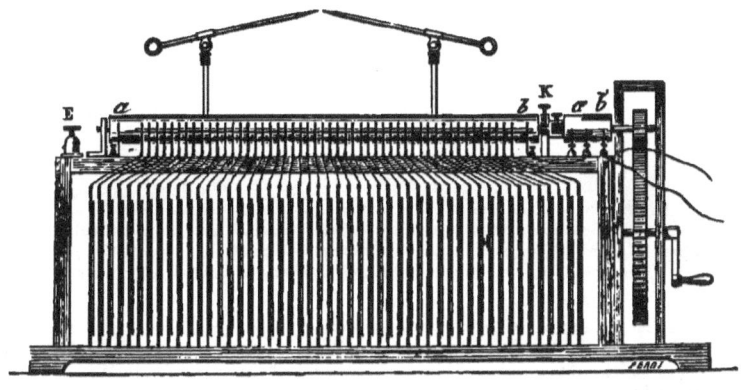

Fig. 87.

When, on the contrary, effects of tension are wanted, the button K must be loosened and the next one pressed which unites the axis of the two cylinders $a\ b\ a'\ b'$, and the wires from the battery must be connected with the terminals of the long commutator. The two cylinders turn together; but the shorter does not fulfil any function, the first unites the condensers successively in parallel and in series.

304. If one sought to obtain only effects of quantity, the cylinder $a\ b$ should be taken off and the machine reduced to the small commutator $a'\ b'$ under which should be placed a vertical or horizontal battery of condensers connected in parallel.

305. **Static sparks of quantity.**—By putting the rheostatic machine, arranged as we have just described, in communication with a secondary battery of from 400 to 800 couples[1], and by rapidly rotating the little commutator, a continued series of noisy sparks is obtained, but very short ($\frac{1}{10}$ to $\frac{1}{20}$ of a millimetre), presenting the appearance of a very brilliant point surrounded with a halo and throwing forth rays of particles from the electrodes.

This kind of spark, though in some respects similar to that of induction, has a character peculiar to itself and produces different effects.

---

(1) In order to facilitate the means of trying experiments that can be made with the rheostatic machine for quantity or tension, we here indicate a simple means of charging a battery of secondary couples designed to work the machine without the necessity of using special batteries with commutators composed of numerous springs for charging and discharging.

These couples may be reduced as we have already explained (vol. 1, note 273), to small thin plates of lead a few millimetres only in breadth. They should be bent in the shape of forks, or tuning forks, keeping a certain length in the upper part to be able to take hold of them easily.

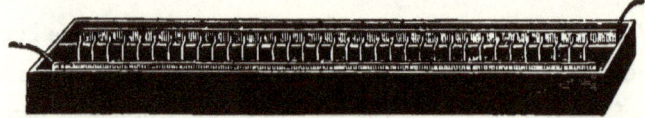

Fig. 88.

A hundred of these little couples are charged at once by placing them in a long narrow tray made of gutta percha, in which there are two compartments about 50 centimetres in length. The couples are placed outside on the separating partition as shown in fig. 88 where, for the purpose of explaining more clearly, only a few are represented. Two long thin plates of lead act as corresponding electrodes which are connected with four Bunsen or six Daniell elements. The electro motive force necessary to charge this system ought to be, in fact, double that sufficient to charge a single secondary cell.

The small plates are charged, if care be taken to *form* them as we have pointed out (53), by changing the direction of the current several times and they end by preserving a large quantity of the charge received. Directly they are charged they should be plunged into closed glass tubes each very close to the other, and the little secondary couples thus formed suffice for making numerous experiments with the rheostatic machine.

The halo is much more developed, especially at the upper part of the luminous point, and is visible without the necessity of insufflation. It forms a crown from 8 to 10 millimetres in diameter.

306. In spite of their static origin and the sharp noise they make, these sparks are of a less tension than those from the secondary battery itself.

Their length is, in fact, less than that of a spark produced directly through the air by a secondary battery of 800 couples (140). Moreover, they do not illumine vacuum tubes, as does the current coming straight from the battery. To obtain light in vacuum with these sparks, the distance between the electrodes must be reduced to one or two millimetres.

307. This inferiority in the tension of the spark of quantity from the rheostatic machine, in comparison with that of the source of electricity which charges it, is explained by comparing the charge of a condenser with that of a voltameter or secondary couple.

The opposing electro motive force yielded by an accumulator of electrical work, whether it be a condenser or secondary cell, could not be greater than that of the electric source itself.

If, in the case of a secondary couple, the principal effect is chemical action on the electrodes and, consequently, the electrolysis of the interposing liquid, there is, in the case of the condenser, a corresponding physical action exercised on the insulating medium which separates them.

From this cause, in both cases, there is a loss of E.M.F. in the current during the charge and an opposing E.M.F. given by the accumulator necessarily inferior to that of the charging current.

308. The greater noise produced by the spark of quantity from the condensers of the rheostatic machine, compared with that made by a spark direct from the secondary battery may be explained in the following manner:

The work effected by a current of dynamic electricity of high tension on a condenser may be considered like putting two opposed surfaces of insulating matter in vibration, reaching to a certain depth depending on the tension of the source or the nature of the substance. This vibration continues a certain time after the charge, like that which would occur by a purely mechanical action on any sonorous substance. From the moment of the discharge, the movement being suddenly counteracted by the speedy return of the matter to its natural state, there results a peculiar noise in the whole of the circuit of discharge, in the spark and even in the insulating matter. Now, one can understand that this sudden decomposition of the compact molecules of a solid substance may be accompanied by a sharp noise, quite different from that resulting from opening or closing the circuit of a high tension battery in which there is no solid insulating substance, but a series of liquid conductors which form the principle part of it.

Thus, the noise of the static spark of quantity, of which we speak, compared with that of the spark direct from the secondary battery, in spite of a lower tension, appears to us to be the result of the particular nature of the discharge, or, more exactly, of the nature of the insulating matter to which the electric vibration has been communicated during the charge.

309. **Calorific effects.**—The calorific power of these sparks of quantity from the rheostatic machine is naturally greater than

that of sparks of tension from the same apparatus. Platinum or steel wires, from 10 to 20 centimetres in length and from $\frac{1}{10}$ to $\frac{1}{40}$ of a millimetre in diameter, may be reddened or melted, whilst the longest sparks of tension would pass along them too easily to produce any perceptible heat.

310. **Mechanical effects.**—The mechanical effects produced are very strong. If these static sparks of quantity be passed through a voltameter filled with a solution of salt, of which the negative pole is a Wollaston electrode, and where the long sparks of tension would run through silently, the passage of the former is accompanied by a very loud sound similar to a small explosion; the mechanical effect produced is so strong that even the vessel of the voltameter is displaced and pushed forward on its stand; the glass begins to vibrate, and, if the commutator be quickly turned, loud rolling or ringing noise is the result.

311. By disposing the connections so that the secondary battery acts at the same time on the voltameter, by the intermediary of imperfect contact, continued breaks are produced spontaneously, and the ringing sound becomes automatic. Any rhythm given to these interruptions is repeated with great intensity in the voltameter and it might be possible, perhaps to make use of this in the telephone.

312. **Rheostatic hydraulic ram.**—The greater part of the phenomena that we have observed in employing currents of high tension, are shown with greater facility and less tendency to be transformed into calorific effects, by the aid of these continuous discharges of semi-dynamic semi-static origin. The experiment we have described by the name of "voltaic pump" (148) is here reproduced very clearly by an altogether mechanical action of electric

force. Instead of rising uninterruptedly, as with a continuous current, the water ascends by jerks, more closely together as the sparks follow each other more quickly, and the apparatus then becomes a true rheostatic hydraulic ram.

By a continual interruption of the current, spontaneously produced as in the preceding case, the effect also becomes automatic.

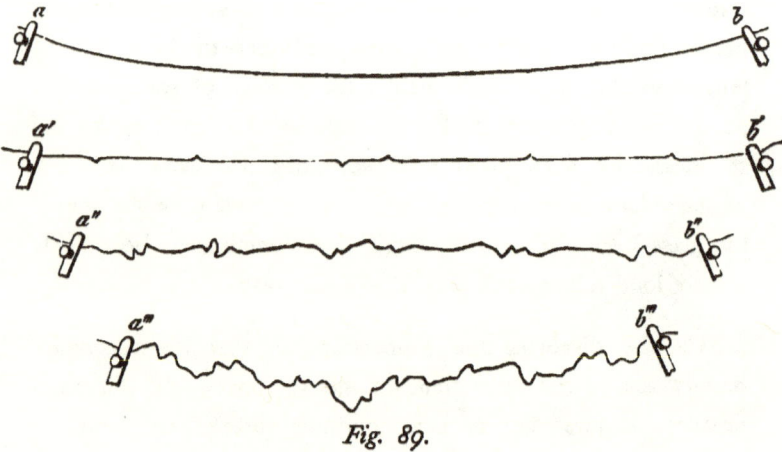

Fig. 89.

313. **Vibration knots formed in a metallic wire by the current of quantity from the rheostatic machine.—** The passage of the current of quantity from the rheostatic machine through very fine platinum wires (of $\frac{1}{10}$ of a millimetre in diameter) is accompanied by remarkable machanical effects.

As soon as the machine is turned, sharp angles appear throughout the length of the wire (about 0^m40), at semi-regular distances, forming a series of accolades or vibration knots. The wire, which was half stretched, rises and changes from the form ab to the form $a'b'$ (fig. 89).

These angles appear to be nearly equal distances apart, but sometimes two or three consecutive ones are seen with the angle pointing in the same direction.

If the machine be again turned, after having placed the nippers in which the wire is held nearer together so that it does not stretch to the point of breaking, new bends appear round the angles already formed and the wire becomes like $a''b''$. If it is shortened to 0^m10 in length it becomes white hot, presenting numerous angles and sinuosities so sharp ($a'''b'''$) that it has the appearance of a continuous electric spark.

In the latter case, it is found shortened after the experiment to the extent of 5 or 6 millimetres in the length of 10 centimetres.[1]

314. There is occasion to remark, that, if the platinum wire is new and annealed by the wire drawer, and if it has not been previously reddened by any kind of calorific source or current, it forms these knots of vibration much less easily. This shows that the current must contend with molecular cohesion to produce this phenomenon.

---

(1) These phenomena may be compared with those which have been observed by Nairne and by M. Edmond Becquerel with discharges from Leyden batteries, and with those observed when a long fine wire is made red by a battery composed of a great number of elements. But here they are more marked and present other characteristics by reason of the different nature of the electric source employed which possesses at the same time the dynamic and static state by the quantity and the tension of the electricity in play.

Nairne had observed that metallic wires submitted to discharges of static electricity underwent a diminution in their length.

M. Edmond Becquerel found that this diminution was inversely proportioned to the cube of the diameter of the wire, and has further observed that the wires become undulated by the action of discharges given by two Leyden batteries of nine jars. The undulations increase in height, according as the discharges follow each other, without ever disappearing to make room for others. (See Traité d'électricité in 3 vol. by MM. Becquerel vol. I, p. 309).

315. **Variations of the distance of the knots according to tension of the current.**—The distances at which these knots or sharp angles appear so clearly from the first instant of the current passing, represented at $a'\,b'$ (fig. 89), do not depend on the rate of rotation of the rheostatic machine. If the rate is great it only makes the transformation of the current from the secondary battery less complete.

But these intervals between the knots appear to depend on the tension of the current. Thus, by diminishing the number of secondary couples working on the rheostatic machine to half; in reducing them from 800 (the number used in the preceding experiments) to 400, so as to have a very decided difference in the tension of the current of discharge given forth from the machine, we have found that the distances between the knots, which were in the first case from 1 to 2 centimetres,[1] vary in the second case from 2 to 3 centimetres.

The amplitude of this kind of longitudinal vibration, produced by this peculiar electric current, seems to increase according as the tension of the current is diminished.

316. **Noise in the wire.**—While this phenomenon is happening, there is heard around the wire a noise or cracking sound similar to that which would be produced by a spark, although in this case the wire has no break in it.

317. This noise in the wire without the intervention of any electro magnetic action is an important point for consideration.

---

(1) These knots are not formed at perfectly equal distances as we have before remarked (313). They generally succeed each other in couples with a shorter interval between as they continue. Thus, after two intervals of nearly two centimetres, there appears an interval of not more than one centimetre and so on. The investigation of the law which governs these divisions of the wire would merit separate study.

It can only be owing to the molecular disturbances resulting from the passage of the current peculiar to the machine, which causes these sudden contractions and distortions in the body of the substances through which it passes.

This phenomenon shows that a corresponding mechanical effect must be produced inside the insulating matter of the condensers, which are here the source of the current which passes along the wire, just as the calorific or chemic effects observed in the exterior current of a battery are but the reproduction of those which occur in the interior of the battery itself. That would then be the cause of the noise or sound made by a condenser at the time of its charge or discharge.

318. While the wire undergoes these vibrations of a longitudinal appearance, the effect of which is permanent and remains visible, it is subjected also to transverse ones of very great size which strongly disturb it.

319. The longitudinal vibrations are a mechanical effect of the current which must not be attributed to its disconnected nature; for, if the current direct from the secondary battery (rendered disconnected by the same commutator acting as simple interrupter) is passed in the same wire no permanent change can be ascertained in the form of the wire.[1]

The transverse vibrations, on the contrary, result from the calorific effect produced by the current being alternately continuous and interrupted; for, if the current from the battery, in an intermittent manner as above, be made to pass along the wire these

---

(1) Care must be taken in this case, by interposing a column of water, to greatly diminish the quantity of the current from the battery without perceptibly weakening its tension, in order to prevent it reddening or melting the wire.

vibrations are seen, although less strongly than with the current from the rheostatic machine. In this latter case, calorific effect is produced very abruptly and stops in like manner, by successive discharges from the condensers. The consequence is that the wire falls while heating, rises while cooling, and violently trembles under the influence of the current.

320. **Fragility of the wire.**—The wire becomes very brittle after the current has passed through it. If the experiment lasts more than two minutes it breaks spontaneously.

This tendency of a wire to become brittle under the influence of an electric current had already been remarked by Peltier and other observers. But, with ordinary currents of dynamic electricity, it was so slight that it was not altogether admitted.[1] Here it is evident by reason of the peculiar nature of the current employed.

321. **Results relative to lightning conductors.**—If the discharges from this machine, by passing through a fine metallic wire, can produce such a change in the molecular structure as to cause it to break spontaneously after some instants, the passage of the currents from the lightning, which combines in a much greater degree electric quantity and tension, must produce similar effects on much thicker conductors such as the rods or iron cords of lightning conductors.

These conductors may then become very brittle and present invisible ruptures, not only after the direct fall of the lightning that they have been able to carry, but also when they have served for a long time for the silent stream of large quantities

---

(1) See "Résumé de l'histoire de l'électricité et du Magnétisme" by MM. Becquerel, p. 237.

of atmospheric electricity. They may even have received a certain number of discharges without there being any appreciable interruption discovered by the aid of electric instruments and yet be in such a state of molecular fragility that another powerful discharge may achieve the rupture of the conductor, as in the experiments above described.

Thus, accidents may be accounted for which happen with lightning conductors apparently quite sound.[1]

322. **Conclusions drawn from the preceding phenomena relative to the mode of propagation of electricity.**—The phenomena we have just described (313 to 320) are of a nature to throw some light on the mode of propagation of electricity. The molecular vibrations revealed by knots formed in a metallic wire, by the curious noise and by a notable change in its cohesion under the influence of the passage of the dynamo-static current which we have just studied, must be produced in a lesser degree in conducting substances traversed by electric currents of very low tension. This vibration may be too feeble to be perceptible but it is not the less real.

We are then able to conclude that the electric movement must diffuse itself in substances after the manner of a purely mechanical motion, by a series of very rapid vibrations of the more or less elastic matter through which it passes.

These facts may be compared with those we have observed when a current of high tension passes above the surface of water,

---

[1] It may be necessary to completely renew a lightning conductor, even though it seem to possess sufficient conductibility, if it be often exposed to frequent violent storms so as to become brittle in consequence of the great quantity of electricity which passes through it. M. Callaud has often had occasion to observe that the cables of lightning conductors become very brittle, and has attributed this fact to the passage of electricity. (Traité des paratonnerres, par A. Callaud, p. 91).

and this, being in vibration, presents to view a number of remarkable luminous forms which recall to mind those made by vibrating plates in acoustic experiments (138).

In these phenomena are found novel analogies between electric and vibrating sonorous motion which is itself a mechanical motion of ponderable matter.[1]

323. **On the quality of reversal possessed by the rheostatic machine.**—If, instead of passing a current of dynamic electricity into the rheostatic machine to obtain static electric effects, the machine be put in connection with a direct source of static electricity, such as an electric machine or another rheostatic machine at work, tokens of reverse transformation are obtained, that is, traces of dynamic electricity.

In this case, the tension poles of the rheostatic machine are put in connection with the static electric source and the charging poles, which are joined to the battery, communicate with a galvanometer.

If the commutator of the rheostatic machine be at rest in such a position that all the condensers are in series, the electric machine charges them in tension, or in series, even though they may be a large number, because of the thinness of the mica plates. When the commutator is revolved, the needle of the galvanometer marks with an abrupt movement the direction of the dynamic current produced by the discharge of all the condensers combined

---

[1] We had already been led to these conclusions after observing effects produced by electric currents of high tension and we had also considered electric discharge as a mechanical movement; but more particularly as an extremely rapid transporting movement of a very small quantity of animated ponderable matter. (Comptes rendus, t. LXXXVIII, p. 442. Extrait d'un pli déposé à l'Académie des sciences, June 11th, 1877 and opened March 3rd, 1879).

in quantity. This collection of condensers, with very thin insulating plates possessing a large surface, plays the part of a voltaic couple which would nevertheless possess a powerful resistance, because of the nature of the medium which separates the electrodes.

If an electric and a rheostatic machine are set going at the same time, the electric machine will not charge the condensers fast enough for a perceptible deviation of the needle to be observed; but the introduction into the circuit of a telephone, as galvanometer, reveals by a rustling noise the passage of a broken current of small intensity.

In the case of two rheostatic machines being used, one as static electric source, the other as receiver, to obtain an inverse result the effects are more marked.

The rheostatic machine may then be considered as reversible, like most of the machines purposed for the transformation of force. But the generators of static electricity furnish so weak a quantity that, even supposing it to be completely transformed into dynamic electricity, the current obtained would have too little intensity to be advantageously utilised. We restrict ourselves to mentioning these results as giving another example of the bonds which exist between the different modes of manifesting electric motion and to show the possibility of transforming one into the other by the most varied means.[1]

---

[1] Paragraphs 275 to 323, inclusive, are taken from a note published on the 14th July, 1879, (Comptes rendus, t. LXXXIX, p. 76 à 80), and from a pamphlet presented to the Académie des sciences, the 6th Oct., 1879. (Comptes rendus, t. LXXXIX, p. 605)

Sixth Part.

Analogy between electric phenomena and effects produced by mechanical actions. Results relating to the nature of electricity.

> "Convertenda plane est opera ad inquirendas et notandas rerum similitudines et analoga, tam in integralibus quam partibus: illæ enim sunt, quæ naturam uniunt, et constituere scientias incipiunt." (Bacon, *Novum Organum*, lib. II, § 27).

324. Since the first observations we made, by subjecting substances endowed with great molecular mobility, such as liquids,[1] to the action of electric currents of high tension, we have been struck with the analogy which the phenomena produced presented to those which result from the action of mechanical force, correctly speaking, on the same substances, particularly when this force is represented by an extreme rapidity of motion communicated to a small mass of matter.[2]

---

[1] Comptes rendus, May 5th and July 26th, 1875.

[2] We have already pointed out a certain number of these analogies in the course of our researches; but we now think it useful to collect them and draw from them some conclusions.

The relations which exist between the two classes of phenomena have appeared to us more visible in these experiments than in those by static electricity because we put in play a greater quantity of electrified matter; and more evident also than with ordinary electric currents because we employed a much higher tension.

But when once these analogies are discovered, they are easily recognized in a more or less marked degree in nearly all phenomena produced by static or dynamic electricity.

325. If, among the phenomena that we have described some are chosen in which these analogies appear most striking, in the first place may be cited the phenomenon of the luminous waves, produced in the heart of the liquid around the extremity of an electrode pressing against the side of a voltameter, through which an electric current of high tension passes out (157).

The violent movement imparted to the liquid, being stopped by the sides of the voltameter, raises the glass to a temperature high enough for the luminous circles to be formed,[1] and, when the current is of sufficient quantity and tension, luminous waves are formed in the midst of the liquid itself.

There is then found in these phenomena a representation of the waves formed on the surface of a liquid by the fall of a mass of solid matter, modified only by the production of calorific and luminous effects.

The varied luminous shapes, produced by a current of yet greater tension striking the surface of a liquid (138), are quite

---

[1] In this experiment the waves produced remain imprinted on the glass in the form of concentric rings.

analogous, in form, to those which are caused by the fall of liquid drops on the surface of a liquid. If the figures we have observed be compared with those which have been obtained by Mr. Worthington and others[1] there will be found among them some which are nearly identical. We have also mentioned the analogy of these shapes with those which result from the vibration of sonorous plates (II, 322).

We recall to mind that the stratifications of electric light in vacuum, observed by MM. Abria, Grove, &c., which present very decided curves in receivers of sufficient diameter, have been already compared by Gassiot, de la Rive, &c., with waves produced by mechanically shaking a liquid.

326. The phenomenon of the sheaf of finely divided water, produced by an electric current of high tension (143), has its analogue in the mechanical sub-division of a liquid by the action of a jet of condensed air only acting at once on a very small portion of its surface *(Appareils pulverisateurs)*.

327. The phenomena of suction, produced by the flow of an electric current of high tension (pompe voltaïque, bélier rhéostatique) are analogous to those which result from the passage, in a narrow tube, of a liquid or jet of steam impelled at a great speed. (Tube de Venturi, Giffard's injector.[2])

---

(1) Proceedings of the Royal Society, XXV, p. 261, (1876).—La Nature, 5th year.—2nd semestre, p. 236, September, 1877.

(2) There is also a close analogy between these phenomena and the effects of suction recently discovered by M. D. Colladon in waterfalls :—" There may be perceived little sheaves, composed of millions of liquid pearls impelled at a rapidity of motion absolutely incredible, in a contrary direction to the water of the cascade, and quickly ascending towards the summit." (Comptes rendus, vol. LXXXIX, p. 286.—Les Mondes, September 25th, 1879, p. 147).

328. The mechanical, calorific and chemic action produced at the same time by an electric current of a certain tension on the surface of glass, which has led us to the "engraving on glass by electricity" (161), may be compared with the action exercised on this substance by an exceedingly fine jet of sand shot forth under strong pressure, which has been employed for several years in America for engraving on glass.

329. The bubbles, blisters or luminous beads formed along a column of matter through which passes a strong electric current at the time of fusing a wire (§ 83, fig. 23), or the incandescence of a narrow stream of rarefied air by a powerful discharge of atmospheric electricity (188, Eclairs en chapelet), offer a close resemblance to the phenomena which accompany the flow of a liquid through a narrow opening under considerable pressure.

330. The vibration knots formed in a wire by the electric current, either dynamic or static, that we have previously studied (313), show, as we have already explained, the analogy which exists between the electric current and the vibratory sonorous motion caused by a purely mechanical action.

331. The experiment described (271), which consists in exhausting a tube of rarefied air, so as to promote in it a luminous stream of electricity, by increasing the tension of the current of the secondary battery by the momentary addition of that from the rheostatic machine, is analogous to the effect well known in the laws of fluids, which consists in exhausting a syphon and promoting the flow of the liquid by suction.

This analogy may be rendered still more striking. Simply bring a stick of electrified resin or ebonite near to the tube and suddenly take it away so that the light immediately appears in the tube. There is thus produced, at the extremity of one of the electrodes, a kind of suction which, added to the already high tension of the current causes the appearance of a luminous stream.

When this phenomenon is more closely studied, the effect is found to be, principally, at the positive electrode. With a substance, on the contrary, charged with positive electricity, the illumination is caused by approaching the negative electrode. In an electrified substance possessing a rather large surface this effect is very strong; and we have thus been able to cause the continuous illumination of a Giesler tube by a secondary battery of 700 couples, by bringing the metallic plate of an electrophorus to the distance 1^m50 from it.[1]

332. If the best known effects of static or dynamic electricity be considered from this point of view, numerous analogies are found with effects produced by mechanical force, especially when they are compared, as we have done, with mechanical actions in which velocity plays a more important part than the mass of matter in motion.

Thus, there is a great similarity between perforations produced by electricity and those produced by projectiles impelled at a high velocity; also between calorific effects obtained by electricity and those resulting from (so called) mechanical impact, and between gyratory reaction motions produced by a flow of electricity (electric whirls, &c.,) and those worked by water, steam or condensed gas (turbines, &c.)

---

(1) These phenomena may be compared with those observed by William Spottiswoode and Fletcher Moulton in their study of the sensitive state of electric discharges through rarefied gases where numerous examples are found of induction effects on electric light in vacuum. (Philosophical Transactions, 1871, first part, page 165.)

333. Effects are obtained by mechanical means, from the division of matter reduced to its first principles, similar to effects of the same kind caused by electricity, in employing, as force, substances impelled at a high velocity.

A jet of steam projected under strong pressure against the slag of blast furnaces, divides it into numberless threads, forming it into a kind of mineral wool. In the same way, matter impelled by electric movement sub-divides, to an infinite extent, all other matter it finds in its way.

334. Cutting tempered steel with an iron disc has latterly been achieved;[1] now, during this operation "it throws out a continuous jet of sparks and particles of steel apparently at white heat; nevertheless, the hand may be passed through this jet with impunity and a sheet of paper interposed is neither burnt nor blackened. These particles appear to be in a spheroidal state; when cold they take the form of an elongated cone resembling stalagmites; the steel has been really melted."

In these mechanical effects new analogies are found with the mechanical, calorific and even physiological phenomena of electricity. The ordinary electric spark, in spite of its high temperature, does not burn, by reason of the small quantity of ponderable matter in play; the steel also in this case, though melted, does not burn because of its extreme state of sub-division and, in consequence, its growing cold so rapidly.

335. The phenomena of attraction and repulsion which seem so characteristic of electricity can be imitated with the aid of a

---

[1] This result has been obtained by the aid of a machine constructed by M. Jacob Reese. See Bulletin de l'Association scientifique de France, November 6th, 1876, p. 77.)

strongly compressed jet of air escaping through an extremely narrow opening. Balls of different substances, even metal, may be held in equilibrium, attracted or repelled, by a jet of air at high pressure, according to their distance from the opening, density, &c.[1]

The recent works of M. Bjerknes have shown the possibility of also obtaining, by other purely mechanical means, attractions and repulsions similar to those caused by electricity.[2]

MM. Dvorak and Mayer have observed, on the other hand, peculiar phenomena of repulsion at the approach of bodies in vibration.[3]

336. Conclusions relating to the nature of electricity.—The analogies which we have just enumerated permit us, we think, to consider electricity as a purely mechanical motion of ponderable matter.

This movement consists in the extremely rapid flow, or transport, of a very small quantity of matter, in regard to the electric spark, the voltaic arc or electrical discharge in general.

337. Electric motion may occasion gyratory movements the same as mechanical motion—correctly speaking—by an effect from reaction due to the flowing of matter, however small the quantity may be, which escapes from electrified substances.[4]

---

(1) These experiments were repeated at the International Exhibition in 1878 in the American section by the aid of reservoirs of condensed air from the Westinghouse brakes.

(2) Comptes rendus, LXXXVIII, p. 165 and 280; LXXXIX, p. 134; 1879.

(3) Philosophical Magazine, 5th series, VI. p. 225, and Journal de Physique, VIII, p. 25; 1879.

(4) This matter is not electric matter as it was formerly believed to be, but electrified matter borrowed both from the substance itself from which it detaches itself and from the centre through which it passes.

338. Electrical motion may become vibratory like mechanical motion when the ponderable matter resulting from the discharge, coming in contact with a substance of a peculiar elasticity, permits it to transfer the shock received through its entire mass.

This peculiar elasticity constitutes electric conductibility. There is not then any transport of ponderable matter throughout the length of the conducting substance, but diffusion by vibrations similar to those of the sonorous motion or the movement transferred to a series of elastic balls. The phenomenon of the jet of ponderable matter may be also produced at the extremity of the conductor when there is a break in the middle of the wire.

339. This transformation into vibratory motion may take place, to a certain degree, in the electric discharge itself, through an imperfectly conducting medium such as ordinary or rarefied air. There is then both transport and vibratory motion.[1]

340. The very rapid movement of ponderable matter, which constitutes electric discharge, produces as we have before said (327), like the rapid motion of a fluid, a suction or inverse motion in the particles of matter which receive the electric shock, or of that which forms the centre of the matter traversed by the discharge.

From that cause a double movement occurs in two different directions; consequently, a double transport of ponderable matter. To this double movement are due the effects produced in electric discharge which are, by general consent, called positive and negative electricity. Instead of these expressions, which seem to infer two sorts of electricity, the terms "direct electric motion" and "inverse electric motion" may be substituted.

---

(1) It is this double effect which often gives to electric phenomena such complicated appearances.

341. As to the phenomena produced by electricity called static, we consider them as due to the vibratory state of the molecules at the surface of electrified substances, accompanied by a more or less abundant emission of material particles detached from this surface, according to the conditions in which the electrified substances are placed in reference to the surrounding medium.[1]

The phenomenon of the aigrette is a characteristic manifestation of this emission of ponderable matter. The aigrette is always produced in a greater or less degree on different points of a strongly electrified substance; the least wrinkle in the surface will occasion it. This phenomenon then reveals the state of continual discharge in which a substance may be when charged with static electricity.

It may be also said that this emission becomes more evident the nearer the electrified substance chances to be to another substance, not electrified, which serves in some degree as target for the projectiles formed by the molecules from the electrified substance.[2]

342. To sum up in a few words the views herein stated; we think that electricity may be considered as a movement of ponderable matter—a movement of transport given to a very small mass of matter, impelled to an extreme velocity, when there is question

---

(1) The earlier electricians, principally Boyle, Hankshee, &c., had already allowed that material effluvium escapes from electrified substances. This idea appears to us to be still correct at the present time by adding to it the vibratory molecular motion of the surface of these substances.

(2) We have several times had occasion to point out, in our examination of currents of high tension, calorific and luminous effects resulting from these molecular shocks. The experiments made by Mr. Crookes also present numerous and brilliant examples of this kind.

of electric discharge—and a very rapid vibration of the molecules of matter when touching its transmission to a distance in a dynamic form, or its manifestation in a static form on the surface of substances. (Oct. 16th, 1879.)

INDEX.

FIRST PART.

THE ACCUMULATION AND TRANSFORMATION OF THE ENERGY OF THE VOLTAIC BATTERY BY MEANS OF SECONDARY CURRENTS.

CHAPTER I.

The study of secondary currents.

	Page.
Historical.—Secondary currents.—Voltaic polarisation	1
Study of secondary currents produced by various voltameters.—Apparatus used	5
General results of the above	7
Voltameter with copper electrodes and water acidulated by sulphuric acid	8
Voltameter with silver electrodes ...	12
Voltameter with tin electrodes	14
Voltameter with lead electrodes ...	14
Voltameter with aluminium electrodes	18
Voltameter of iron and zinc	19
Voltameter with gold electrodes ...	20
Voltameter of platinum electrodes	21
Voltameters with a saturated solution of bichromate of potash	24
Conclusions ... ,	26

CHAPTER II.

Storage of the energy of the voltaic battery by means of secondary cells with lead plates.

	Page.
Deductions drawn from the preceding researches	29
Secondary cell with coiled lead plates	30
Secondary battery of large surface for quantity	31
Secondary cells with parallel lead plates	31
Another form of secondary cell with lead plates	34
Chemical actions produced in secondary cells with lead plates...	38
Formation in electro-chemical preparation of secondary cell with lead plates	41
Absorbtion of the gases during the charging of secondary cells	46
Maintenance of secondary cells	48
Effects produced by secondary cells with lead plates...	48
Calorific effects...	49
Magnetic effects	50
Formation of ozone in secondary cells and voltameters with lead plates	51
Shades or tints of oxide produced at the positive pole during the discharge of secondary cells	53
Duration of the discharge in secondary cells	54
Constancy in the secondary current during the discharge	55
Preservation of the charge taken by secondary cells...	57
Residual charge afforded by secondary cells	59
Increase of the intensity of a secondary cell with rest after being charged	60
The electro-motive force of lead plate secondary cells	61
Resistance of lead plate secondary cells	63
Necessary E.M.F. of the primary current for charging the secondary cells	64
Limit of charge that secondary cells may take	65
Secondary cell charged by a thermopile	66
Secondary cell charged and discharged by means of the Gramme machine	66
Various analogies presented by secondary cells	68
Return obtained from secondary cells	70

CHAPTER III.

Transformation of the energy of the voltaic battery by means of lead plate secondary batteries.

	Page.
Historical.—Various means used in order to increase the E.M.F. of the battery	73
Secondary battery for tension effects formed of parallel lead plates ...	75
Secondary battery for currents of high tension formed of couples composed of coiled lead plates	77
Effects produced by secondary batteries composed of lead plates ...	82
Large secondary battery of extended surface for effects of quantity or tension	83
Secondary battery of lead plates for prolonged effects of tension ...	84
Instructions respecting the use of secondary batteries	85
Returns.—Comparisons	88

SECOND PART.

APPLICATIONS.

Galvanocaustic applications	90
Use in lighting dark cavities in the human body, and obscure hollows in general	94
Application to firing mines, etc.	95
Application to domestic uses.—Saturn's "tinder box"	97
Application to electric breaks for use on railways	102
Application to the eudiometric analysis of the atmosphere of mines ...	103
Application to the production of luminous signals	103
Application to the production of the electric light in special cases ...	104
Application to the sub-division of the electric light	105
Physiological effects produced by secondary batteries	106
Various applications	106

THIRD PART.

EFFECTS PRODUCED BY ELECTRIC CURRENTS OF HIGH TENSION.

CHAPTER I.

	Page.
Use of secondary batteries for the study of these effects	108
Experiment upon the "luminous sheath" with the current decreasing in intensity	110
Change of colour in the "luminous sheath" according to the intensity of the current	111
Batteries of from two hundred to eight hundred secondary cells, used in studying electrical effects of high tension	113
Luminous liquid globules	117
Globular discharge.—Brush discharge and luminous figures produced by the discharge of a battery of 800 secondary couples	120
Wandering electric sparks	123
Sheaf of aqueous globules	126
Jets of vapour	127
Electrified liquid vein—gyratory motion	128
Electric bar	130
Voltaic pump	131
Detonations produced at the extremity of the positive electrode	133
Electro-silicious light	135
Crowns, arcs, rays and undulating motions	139
Electro dynamic whirls	140
Crater-like perforations	143

CHAPTER II.

Engraving on glass by elctricity	146
Electric boring or drilling	149
Different applications	150

265

FOURTH PART.

CHAPTER I.

	Page.
Analogies with globular lightning...	151
On the formation of globular lightning	152
On the light emitted by globular lightning ...	154
On the noise accompanying globular lightning	156
On its gyratory motion ...	156
On its immediate disappearance ...	157
On the slow passage of globular lightning ...	157
On the noise accompanying the appearance and disappearance of globular lightning ...	160
On the action of many pointed lightning conductors in cases of globular lightning ...	162
Instance of globular lightning at Paris in 1876	163
Lightning "en chapelet" ...	165
On the formation of lightning and its relations with globular lightning ...	169

CHAPTER II.

Comparisons with the phenomenon of hail.

On the formation of hail ...	173
On the mechanical and calorific effects of electricity in the formation of hail ...	174
On the wind, noise, lightning—with or without thunder—accompanying hail ...	176
On hail produced without any apparent electrical manifestation	177
On the short duration of falls of hail ...	177
On the tracks of hail and rain ...	177
On the intermittences and returning forces of hail storms	178
On the form and glare of hailstones ...	179
On their internal structure and their size ...	179
On the whirlwinds accompanying hailstorms and the cause of their spiral motion ...	181
Conclusions ...	182

CHAPTER III.

Comparisons with waterspouts.

	Page.
On the luminous effects, noise, jets of vapour, &c., accompanying waterspouts	183
On the spiral motion of whirlwinds and cyclones	184
On the spiral motion of simoons	184
On the spiral motion of lightning...	186
On the waterspouts produced without apparent electrical phenomena ...	186
On the electrical signs of waterspouts	186
On the cause of the descent of waterspouts	186
On the comparison of tidal waves and "bores" (seiches) with the "electric bore" experiment	187
On the phenomena of the suction of waterspouts and their analogy with the voltaic pump experiment	187
Conclusions	188

CHAPTER IV.

Comparisons with the polar aurora.

On the luminous effects, crowns and arcs and their undulating motion ...	189
On the dark segment or circle of the polar aurora	190
On the fluctuation of the light	191
On the falls of rain and snow, and the wind accompanying the polar aurora	191
On the noise accompanying the polar aurora	192
On the magnetic disturbances	192
On the electrical signs of the polar aurora	192
On the cause of the discharge of electricity in the polar aurora ...	193
On the origin of atmospheric electricity	193
Conclusion	197

CHAPTER V.

Comparisons with spiral nebulæ	198

www.ingramcontent.com/pod-product-compliance
Lightning Source LLC
Chambersburg PA
CBHW031940230426
43672CB00010B/1999